Functional Morphology of the Vertebrate Respiratory Systems

Biological Systems in Vertebrates

Series Editors
H.M. Dutta and Douglas W. Kline
Kent State University
Kent, Ohio, USA

Volume I: Functional Morphology of the Vertebrate Respiratory Systems

Functional Morphology of the Vertebrate Respiratory Systems

J.N. Maina

School of Anatomical Sciences
University of the Witwatersrand
Johannesburg
South Africa

CRC Press
Taylor & Francis Group
Boca Raton London New York

CRC Press is an imprint of the
Taylor & Francis Group, an **informa** business

A SCIENCE PUBLISHERS BOOK

First published 2002 by Science Publishers, Inc.

Published 2019 by CRC Press
Taylor & Francis Group
6000 Broken Sound Parkway NW, Suite 300
Boca Raton, FL 33487-2742

ISBN-13: 978-1-57808-252-0 (pbk)
ISBN-13: 978-1-57808-253-7 (hbk)

Visit the Taylor & Francis Web site at
http://www.taylorandfrancis.com

and the CRC Press Web site at
http://www.crcpress.com

CIP data will be provided on request.

DEDICATION

To Wanjuku, Ndegwa, Wanjiru, and Kireru for their absolute support.

Preface

While the actual writing and preparation of this book started lately, its conception started more than ten years ago. After using various morphological techniques to study respiratory functional morphology from a comparative approach, I contemplated preparing a book that would be easy to follow and understand. As much as possible, such work would avoid many unfamiliar terms and "interrupting" references in the text. Inasmuch as that might appear to depart from the established norm in scientific communication, I found the challenge attractive and worth pursuing for three main reasons. Firstly, I have enjoyed the challenges and the fascinating findings that I continue to encounter as I seek to understand the myriad strategies and forms of structures that different animals have inaugurated for acquisition of molecular oxygen. It gives me great pleasure to have this opportunity to share the "story". Secondly, from a personal perspective, without wanting to appear unfairly critical, in science we more or less write for ourselves or for an exclusive circle of experts. Few books attract broad readership outside the realm of persons proficient in the particular subject. In fact, if we go by readership, we have a thing or two to learn (in the art of writing) from science fiction writers who have built a dedicated, ever-increasing readership. Thirdly, it is incumbent upon us (for the sake of future and progressive science) to draw as broad a readership as possible into our circle of interests and activities. This will partly be achieved if we simplify and improve our art and method of communication. I am convinced that we "sell ourselves short" if and when we fail to take the initiative to do so. In most countries and important international forums, nonscientists are considerably involved in making and executing important decisions that profoundly impress upon science. Areas such as environmental conservation and management, genomic studies, and use of animals in research and experimentation in particular stand out. Unequivocally, a well-informed person is likely to make a correct decision. In seeking to achieve my objective, I have intentionally prepared a book with less text and more illustrative material. I expect that the impact of the differences and similarities in the design and construction of the evolved respiratory structures will be more forcefully communicated visually. I will leave it to the reader to judge whether the goal has been accomplished. For sagacious readers, I have included a comprehensive list of references used herein and further reading material at the end of the book.

The field of comparative respiratory morphology has recently started to attract great interest as it embraces new techniques, especially molecular biology. Most of the published work in the area is scattered, however, in biological, zoological, microscopical, morphological, and physiological books and journals. Apart from two recent publications[1, 2], reference work in the form of books, treatises, monographs, and theme journals, are presently lacking on the subject. It is hoped that this book will provide a reference base. Chapter one outlines the fundamental factors that prescribed the design of the gas exchangers and the principles upon which the constructions were founded. Subsequent chapters are purposely sequenced to survey the progressive developments in the evolution of vertebrate respiratory organs, more or less in the order that they are chronologically perceived to have occurred.

This book was written with a broad readership in mind. Students of biology as well as experts in areas of zoology, physiology, morphology, biological microscopy, biomedical engineering, paleontology, ecology, and those persons already involved in or contemplating the study of materials and aspects relating to respiration in whole organisms will find it useful. Scientists in Earth sciences keen in the consequences of past intercourse between environmental factors (the physical realm) and evolution as well as in adaptation (biological domain)—an interaction upon which the composition and patterning of the extant animal life has sprung—will find the book of interest. To explicate the essence of the designs of respiratory organs in vertebrates together with the shared and various stratagems adopted for acquisition of molecular oxygen, an ecological-paleobiological-evolutionary approach was adopted in preparing this book. Past conditions, states, events, and circumstances (factors that bear strongly on the present) have been analyzed and used to help elucidate when, how, and why certain vertebrate respiratory structures were inaugurated and others avoided. Of all the concepts enunciated in biology, that of evolution is probably the most important and encompassing. Dobzhansky[3] declared that "nothing in biology makes sense, except in right of evolution". As will be evident in this account, I totally embrace that view. In order to give a truly comparative account, I have considered broad taxonomic groups of animals. Aspects concerning individual species are only given to illustrate a particular theme, principle, or point. It is pertinent to mention that the word "morphology" is purposely used in the title of this book and frequently in the text. Morphology is not a synonym for "anatomy", i.e., description of parts of a biological entity for its own sake, but unfortunately has been so loosely used that today the two terms are interchanged. Morphology is a discipline that delves into the attributes and organization of a structure for the purpose of explicating their logical basis. D'Arcy Thompson[4] understood the discipline as follows: "Morphology is not only a study of material things and of the forms of material things, but has its dynamical aspect, under which we deal with the interpretation (in terms of force) of the operations of energy." That view reflects the true spirit of the discipline, correctly denoting the etymology of the word. The word "design[5]" and the term "gas exchanger[6]" are frequently used in the text and are defined in footnotes below.

J.N. Maina (DVSc)
Johannesburg, February 2002.

---

[1]Maina, J. N. 1998. The Gas Exchangers: Structure, Function and Evolution of Respiratory Processes. Springer-Verlag, Berlin-Heidelberg.

[2]Maina, J. N. 2002. Fundamental Aspects and Features in the Design of Gas Exchangers: Comparative Perspectives. Springer-Verlag, Berlin-Heidelberg.

[3]Dobzhansky, T. 1973. Nothing in biology makes sense, except in the right of evolution. *Amer. Biol. Teacher.* 35: 125-129.

[4]Thompson, D'arcy. 1959. On Growth and Form. Cambridge University Press, Cambridge (2nd ed.).

[5]The word "design" is borrowed from engineering. It is used here in the sense "creative arrangement of parts in a respiratory structure brought about by natural selection".

[6]The term "gas exchanger" is used to include all structures that have either primarily evolved for or have secondarily acquired a respiratory function, irrespective of the species in which they occur, their location, structural complexity, and respiratory medium utilized.

Acknowledgments

The work on which this account is based wasn't accomplished single-handedly. Over the years, I have been particularly lucky to collaborate with many colleagues who generously shared ideas and readily gave me their precious time. I deeply appreciate their contribution and more importantly their continuing friendship. Unfortunately, all of them cannot be named here. Special mention must be made, however, of Prof. (Emeritus) A.S. King, University of Liverpool; Dr. M.A. Abdalla, King Saud University; Prof. G.M.O. Maloiy, University of Nairobi; Prof. S.P. Thomas, Duquesne University; and Prof. C.M. Wood, McMaster University. My past and present students have been pivotal in shaping the thoughts expressed here. I must hasten to add, however, that any infelicities of judgment that may have occurred are totally mine. My most sincere thanks to Mr. Moses Mwasela-Tangai for preparing the line drawings used in this book. Some of the photo-and electron micrographs included here have been published before. I acknowledge the permission generously granted by the following publishers to reproduce materials that originally appeared in their publications:

Academic Press Ltd

Birkhähauser-Verlag

Cambridge University Press

Elsevier/North Holland Publishing Company

Lea and Febiger

Longman Group (UK) Ltd

Oxford University Press

Pergamon Press

Springer-Verlag

The Company of Biologists Ltd

Wiley/Liss Inc.

Contents

Preface   vii

Acknowledgements   ıx

1. Respiration—Fundamental Principles and Concepts   **1**

   1. Oxygen: a paradoxical molecule   1

   2. General principles in the engineering of natural and biological structures   3

   3. Quintessential design of the respiratory organs   5

   4. Evagination and invagination: prognostic designs of the respiratory organs   10

   5. Geometry and association of structural components in the respiratory organs   14

   6. Fractal geometry: quintessential design in nature and the respiratory organs   15

   7. Structural-functional correlations in the design of respiratory organs   17

2. Gills   **19**

   1. Evolution of water-breathing and development of gills   19

   2. Structure of fish gills   20

   3. Functional design of fish gills   31

3. Skin   **39**

   1. The skin: is it an archaic or a novel respiratory structure?   39

   2. Versatility of the skin as a gas exchanger   40

4. Swim (Air) Bladder   **46**

   1. Evolution   46

   2. Development   46

   3. Location and structure   47

   4. Function   57

5. Transitional (Bimodal) Breathers   **60**

   1. Evolution of bimodal breathing   60

   2. Diversity and structure of accessory respiratory organs   61

6. Amphibian Lung   70

1. Evolution of amphibians and mutable respiratory modalities   70
2. Structure and heterogeneity of amphibian respiratory organs   71

7. Reptilian Lung   85

1. Evolution of reptiles, air-breathing, and terrestriality   85
2. Quintessential design of reptilian lungs   86
3. Structural heterogeneity of the reptilian lungs   86

8. Avian Lung   101

1. Evolution of birds and the highly efficient parabronchial lung   101
2. Flight: a unique and energy costly mode of locomotion   102
3. Structure of the avian respiratory system   102
4. Quantitative morphology of the avian lung   113
5. Functional design of the avian lung   114
6. Structural-functional correlations in the design of the avian lung 114

9. Mammalian Lung   116

1. Evolution of mammals and the bronchoalveolar lung   116
2. Structure and function of the bat lung   117

10. Summary and Conclusions   131

References and Works to Consult   135
Index   165

1

Respiration—Fundamental Principles and Concepts

1 OXYGEN: A PARADOXICAL MOLECULE

The development of eukaryotic cells from prokaryotic ones (about 2 billion years ago), the realization of sexual reproduction (about 1 billion years ago), and the accretion of independent cells into a cohesive, integrated, multicellular state about 700 million to 1 billion years ago were momentous events in the evolution and progress of animal life. These quantum events culminated in the rise of vertebrates and ultimately that of the endothermic-homeotherms, the most highly metabolically active modern taxon. Few processes in biology are as ancient and as important for life as respiration. While animals can survive for weeks without food and days without water, they continually need molecular oxygen for energy production by oxidative phospholylation. Unlike metabolic substrates such as carbohydrates and fats that can be conserved in large quantities and utilized as needed, oxygen has to be unceasingly contracted from outside. In a person weighing 70 kg, at any one moment there is only about 1.55 L of oxygen in the body. Of the total amount, 370 cm^3 is found in the alveoli, about 280 cm^3 in the arterial blood, 600 cm^3 in the capillary and venous blood, 60 cm^3 dissolved in body tissues, and 240 cm^3 bound to myoglobin. The quantity of oxygen dissolved in the tissues (about 0.8 cm^{-3} kg^{-1}) can support life for about 6 minutes and only for a few seconds during exercise.

Even before the discovery of oxygen by Joseph Priestley in 1771 and three years later determination of the gaseous composition of air by Antoine Lavoisier, it was well known that breathing [i.e., ventilation of the "body" (lungs) with air = pumping air in and out of the "chest" (lungs)], a mechanical process conspicuous particularly in large animals (especially birds and mammals) was essential for life. Until recently, death was loosely associated with cessation of breathing and the common method of killing was by strangulation. The familiar phrase "breath of life" bespeaks the importance of respiration, i.e., acquisition of oxygen, for life. In an adult person, about 12,000 L of air passes through the lungs every day. Respiratory efficiency connotes the speed at which an animal uses its resources to meet the demands placed on it by the environment and the lifestyle that it pursues. Energy generation, its storage, and utilization are processes central to the metabolic performances of animals. Energy drives all biological processes from molecular to ecological levels. It is imperative for maintaining the structural and functional integrity of organisms and fortifying homeostasis against external and internal perturbations. Animals that can achieve and sustain high oxygen to carbon dioxide exchange ratios in relation to their body volumes and those that can establish stable tissue-to-fluid oxygen concentrations under various environmental conditions can attain the highest levels of aerobic metabolism. They are among the most ecologically successful organisms.

Oxygen, a product of photosynthesis by plants and cyanobacteria, has been the most singularly pervasive molecular factor in setting the form, patterning, and geographic distribution of extant animal life. By the start of the Paleozoic (about 600 million years ago), the partial pressure of oxygen (PO_2) in water and air had risen to a modest level of 0.2 kPa [i.e., one-hundredth (= 0.2% oxygen by volume)] of the modern sea level pressure. When the first vertebrates (ostracoderms) appeared on Earth some 550 million years ago, the PO_2 was only 0.9 kPa. In the Devonian Period (some 300 million years ago), when amphibians ventured onto terra firma, the PO_2 had risen to 4.7 kPa. The present level of atmospheric PO_2 (about 21 kPa) was not reached until the Carboniferous Period (some 250 million year ago) when reptiles first appeared on land. The level of atmospheric oxygen shifted greatly in the Phanerozoic. During the Carboniferous Period, for example, it rose to a hyperoxic level of 35% (compared to the present atmospheric level of 20%) and then dropped sharply to a hypoxic low of 15%. These changes were duplicated in water. They had a dramatic effect on aquatic life. The abundance of oxygen during the Mid-Devonian to Carboniferous Periods supported development of exceptionally huge animals such as the giant dragonfly-like Meganeura that reached a body length of 60 cm and a width of 3 cm. The oxygen-rich atmosphere granted higher metabolic capacities to the extant animal life. It motivated radiation into diverse ecological habitats, resulting in unprecedented speciation. Paucity of molecular oxygen during most of the Precambrian is envisaged to have curtailed progress of life from simple unicellular to complex multicellular states. The so-called "Cambrian Explosion," an event epitomized by remarkable speciation, is attributed to an upsurge of molecular oxygen at the Precambrian-Cambrian boundary.

Constituting about one-quarter of the atoms in organic matter, molecular oxygen is a fundamental building block for life. Paradoxically, due to the generation of highly reactive oxidative species, e.g. the superoxide anion radical (O_2^-), hydrogen peroxide (H_2O_2), hydroxyl radical (OH^-), and singlet oxygen ($^1O_2$), molecular oxygen is extremely toxic to carbon-based life. As part of the antioxidant defense system, life has evolved a battery of simple nonenzymatic molecules and complex enzymes that scavenge oxidative oxygen radicals. The former includes glutathione, ascorbate, urate, bilirubin, ubiquinol, β-carotene, and tacopherol, while the latter comprises superoxide dismutase, catalase, and glutathione peroxidase. Superoxide dismutase converts O_2^- to H_2O_2 plus O_2 and catalases and peroxidases convert H_2O_2 to water (H_2O) and oxygen (O_2). Incorporation of an injurious molecule such as oxygen in the biochemistry of energy production indicates the importance of efficient metabolic processes and energy production for animal life. The relatively small molecular weight of oxygen, its high intracellular diffusivity, and appropriate redox potential promoted its utilization as a proton acceptor in the tricarboxylic chain reaction. Moreover, water, the end product of aerobic metabolism, is both an innocuous and a necessary substance for life.

Fermentation (glycolysis) is an inefficient way of producing energy. Much of it is left secured in the chemical bonds of organic molecules such as alcohols and organic acids—harmful products that must be cleared before they accumulate to toxic levels. Through glycolysis, a molecule of glucose yields only 2 molecules of adenosinetriphosphate (ATP) that contain about 15 kCal energy compared to 36 ATP molecules (= 263 kCal energy) produced through aerobic metabolism. When it evolved, aerobic metabolism granted an efficient means of energy production, allowing animals to invest the excess in founding complex, more optimal, and adaptable states.

While there have been assertions that life can exist without oxygen in ordinary habitats, such cases can only occur in the simplest animal life. Intestinal parasites are purported to live without oxygen and intertidal molluscan facultative anaerobes to remain for days without it. In adverse conditions, some animals enter latent (ametabolic) states. Cryptobiosis, a condition wherein life practically stops, is the most extreme of such states. However, even in such conditions an infinitely small quantity of energy is necessary to maintain important processes such as protein turnover and ion flux. In certain unique habitats, life can exist without oxygen. For example, in submarine geothermal plumes occurring at depths of 3 km or more below the ocean surface (sites where oxygen is lacking since photosynthesis is not possible due to lack of sunlight), alternative pathways of energy production have evolved. Chemoautotrophic endosymbiotic bacteria break down hydrogen sulfide (commonly present in abundance), producing ATP. This particular strategy displays nature's profound inventiveness in circumventing the many constraints that eventuate along life's pathways. The hydrogen sulfide/sulfur-based energy production cycle that exists around volcanic submarine vents supports flourishing colonies of animals that include giant pogonophoran tube worms, crabs, shrimps, giant clams, fishes, and mussels. It is worth noting that some of these species were unknown to science 30 years ago!

A comprehensive account of the functional morphology of the vertebrate respiratory organs must delve into how different organs and organ-systems have evolved, developed, been refined, and integrated for the purpose of gas exchange. Moreover, the description must explore these states and phenomena outside the purview of the so-called "model animals". Among the elasmobranchs, the commonly studied species are a variety of dogfish, e.g. *Scyliorhinus canicula*, *Squalus suckleyi*, *Squalus acanthias*, and skates. Among the bony fish (class: Pisces) studies have been made largely on the subclass Teleosti, with the most highly investigated species being the cod (*Gadus morhua*), eel (*Anguilla anguilla*), goldfish (*Carassius auratus*), trout (*Onchorhynchus mykiss*, formerly *Salmo gairdneri*), and the sea raven (*Hemitripterus americanus*). In amphibians, the common grass frog (*Rana pipiens*), European frog (*Rana temporaria*), and the marine toad (*Bufo marinus*), all of which are anurans, are unfortunately taken to be representative of the diverse class Amphibia. Within the class Reptilia, particular interest has been shown in painted turtles either of genus *Pseudemys* or *Chrysemys*. The laboratory white rat (*Rattus rattus*), mouse (*Mus musculus*), and the guinea pig (*Carvia porcellus*) have been widely used among mammals while in birds, the domestic fowl (*Gallus gallus* variant *domesticus*), pigeon (*Columba livia*), muscovy duck (*Cairina moschata*), and guinea fowl (*Numida meleagris*) have been preferred. These few animals, most of which were selected more for convenience and availability than for any particular morphological or physiological merit, are not representative of the large vertebrate taxa. Many contradictory views and conflicts occurring in biology (including comparative respiratory morphology) emanate from unwarranted extrapolations and generalizations based on narrow observations and findings noted for a few unrepresentative animals.

2 GENERAL PRINCIPLES IN THE ENGINEERING OF NATURAL AND BIOLOGICAL STRUCTURES

In biology, the evolution of sound designs has been attained at enormous cost. About 99.99% of all species of animals ever evolved on Earth are now extinct. Regarding their makeup for survival, for all intents and purposes the animals that succumbed may be considered failed experiments. Except for metals, nature has experimented with virtually every construction material and produced practically every device—except the wheel! Parsimonious as nature is, the rationale behind avoiding the wheel is not all that difficult to appreciate. Since moving parts are the most susceptible to failure, engineers themselves have kept them to a minimum! The forms of nature's technological innovations are intellectually engaging and aesthetically pleasant. Complexity pervades all levels of organization of living matter. A virus, the simplest biological entity, on average contains 10^4 atoms, a complex organic molecule contains well over a million atoms, and a cell about 10^{14} atoms. For all their worth, reductionism and all mechanistic approaches in biology attempt to explain natural phenomena by recognizing, isolating, analyzing, and manipulating fewer and simpler components of a complex structure that are responsive to exact simple laws of physics.

While not very often declared, expressly or tacitly, the goal of biological science is to determine the rules that control the workings of cells, organs, organ-systems, and organisms. Biologists believe that their results will eventually be explicable at cellular and molecular levels. As observed by the great Greek philosopher Aristotle (332 B.C.), every structure exists for some reason—although the actual purpose may not always be clear, especially if the structure is a biological one. Nature doesn't yield its secrets willingly: they have to be diligently teased from it through well-planned inquiries and painstaking attention to details. In engineering design, most devices have a single purpose. However, in biology, many structures perform more than one function, either simultaneously or successfully. If choices of structural design are many in engineering, they should be more abundant in biology. The occurrence of various solutions to a given structural problem is one of the reasons why animals have evolved into such a large variety of forms. Life's prolific tree has produced between 5 and 50 million species of animals!

The task of developing an efficient structure is not an easy one. Billions of years of natural selection have provided a cornucopia of exquisite, imperfect, or intermediate biological structures. In spite of the fact that materials found in biology are often very different from those used in engineering, the geometries of the structures in which materials can be used to support loads are fairly much the same. Certain structural states and forms best exploit the strength properties of the materials. Albeit the fact that animals have yet to invent the wheel, nature is generally more clever than engineers at developing the potential of a given structural concept. The engineering of nature is mainly an engineering of soft tissues. Soft, resilient tissues able to support existing loads and grow and evolve have been utilized to a great extent. A common cause of accidents in engineering arises from structural failure resulting from the designer's lack of correctly anticipating the magnitude and the direction of the loads that have to be resisted. Just as in architectural designs, in biology, the structural materials must have definable physical properties such as strength, insulation, and elasticity.

The scale of adjustment that allows a biological system to cope with shifting functional loads constitutes a reserve capacity (= safety factor). Such sufficiencies may be considered "excessive constructions" over and above those essential for basic performance. In engineering schemes, a safety factor is defined as "the ratio between the load that just causes failure of a device, i.e., the component's maximal capacity (= strength) to the maximum load that the device is anticipated to bear during operation". Safety factors vary greatly between different tissues and organs. Biological systems change harmonically with the fluctuating strains and stresses to which they are subjected. In composite structures (such as biological tissues), theoretically there should be room for infinite creativity. Physical (constructional) and biological (phylogenetic, developmental, functional, and ecological) controls prescribe the number of feasible outcomes (= phenotypes). Enhancing the safety margin of operation exacts commitment of more resources for construction and maintenance. By natural selection acting on the phenotype, the values of safety factors in different tissues, organs, and organ-systems is aligned to particular performance needs. Excess capacities and extravagant structures occasion unnecessary costs in terms of energy consumption for infrastructural maintenance and performance.

Unlike human engineers, nature has avoided preexisting inorganic solids, but has elected to evolve, de novo, organic weight-bearing materials suitable to a particular purpose. Biology places great premium on strength and mechanical safety of tissues, while also exacting lightness (in weight) and metabolic efficiency. As one pursues "new" technologies, one is simply getting back to what nature has developed, tested, and refined. Over the about 4 billion years of evolution of life, "errors and corrections" have been made. In what may be considered convergence between natural forms and technology, research into the designs of biological structures has advanced into the exciting scientific discipline of bionics (= biomechanics). The discipline is dedicated to identifying nature's constructions and their possible applications in humankind engineering. Parallel technical principles in the "plans" and "constructions" between the physical (inanimate) and the living (animate) structures occur. For example, to solve certain packaging and mechanical strutting architectural problems for optimal geometric arrangement, one need only make correct selection from among a vast array of models available in nature, without necessarily having to do original research. For instance, in the middle of the 19th century, Sir Joseph Paxton based the design of the Crystal Palace of London on the supports and reinforcements of the water lily *Victoria amazonica*. It is highly probable that the organization of certain biological structures (e.g. in the honeycomb and in the arrangement of siliceous particles of tiny diatoms) inspired the construction of Fuller's geodesic (triangular or hexagonal arrangements of structural elements) self-supporting domes. Obliging economical utilization of constructional material, such designs provide maximal utilizable space while conferring great physical strength. The precision entailed in the construction of the honeycomb of bees is amazing. The cells are built to a thickness of 0.073 mm (with deviations of no more than 2%), the diameter of the individual cells is 6.5 mm (with a 5% tolerance), and the inclination is 13% from the horizontal. During construction of dams by the beavers, a considerable amount of instinctive conceptual "knowledge" of engineering physics, i.e., in hydromechanics and mechanics has to be committed in their elegant fabrications. In all evolved biological structures, the spider's dragline (a tool that determines survival by providing means for procurement of food and escape from predators—a lifeline in the true sense of the word!) perhaps best illustrates the process of evolutionary optimization in biological design. The dragline is a multiphase material that consists of double filaments. Individually the filaments can support the weight of the animal (in case one is accidentally cut!). A single line will break at a stress equivalent to that generated by about 6 times the spider's weight. Activities such as movement, jumping, rapid descent, and rapid ascent (when higher stress is exerted on the dragline) are well accommodated in the design. The elastic limit of a dragline gives the maximum safety (a safety coefficient being the ratio between the mechanical strength of the dragline and the spider's weight) for supporting the arachnid's weight. The mechanical properties of spider draglines have been refined over about 400 million years of evolution of arachnids.

Body size is not a simple vagary of nature. Strict physical laws imposed upon the mechanical structure of an organism, the environment in which it subsists, and the life styles pursued determine it. Gravity is one of the factors that have set the sizes of animals. Regarding makeup and performance of living things, the laws of physics set the boundary conditions that cannot be transgressed. However, the laws themselves do not usually directly prescribe the actual mechanisms employed to circumvent or overcome the constraints imposed by natural selection. In the processes of evolution and adaptation, animals are not passive participants totally subservient to the drifts of selective pressures: they carefully identify the course they wish to take, actively engage the pressures, and set the momentum as well as the nature of change.

3 QUINTESSENTIAL DESIGN OF THE RESPIRATORY ORGANS

Respiration, the process in which molecular oxygen is appropriated from an external fluid medium and channeled to the tissue cells for energy production, incorporates a complex, highly integrated arsenal of structural components and functional processes. Together, biomechanical, physiological, and behavioral processes participate in making a sample of the external respiratory medium available to an organism from which molecular oxygen is extracted and carbon dioxide voided into it. External respiration involves the movement of two vectorial quantities—influx and efflux of oxygen and carbon dioxide respectively. Through convection and diffusion, oxygen is delivered to the tissue cells across an assemblage of tissue compartments [Figs. 1-4]. Regarding their structure and function, at the most basic level of organization the respiratory organs are markedly similar. A thin tissue barrier separates external (water/air) and internal (blood) respiratory media. Moreover, a partial

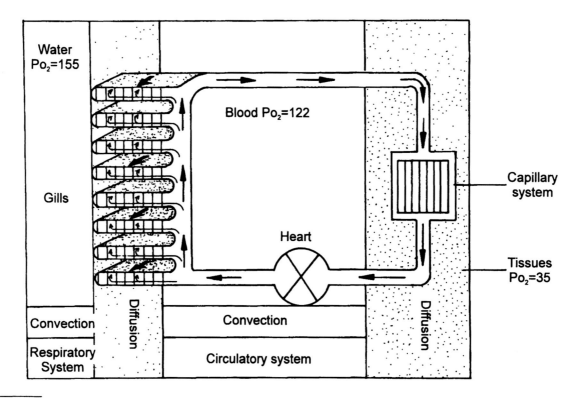

Fig. 1 Integration of the respiratory organs (gills) and the circulatory system in the transfer of oxygen from the external medium (water) to the body tissue cells. Sites of convective and diffusive gas transfer are shown. In gills, water and blood are exposed to each other in a countercurrent manner (modified after Satchell 1971).

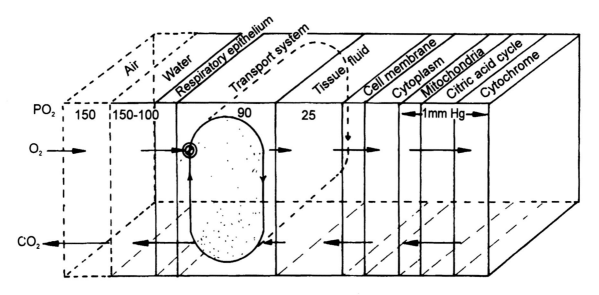

Fig. 2 Influx of oxygen from the external respiratory medium to the mitochondria and efflux of carbon dioxide from the tissue cells. The partial pressure gradient of oxygen decreases down the stratified pathway (modified after Hughes 1978).

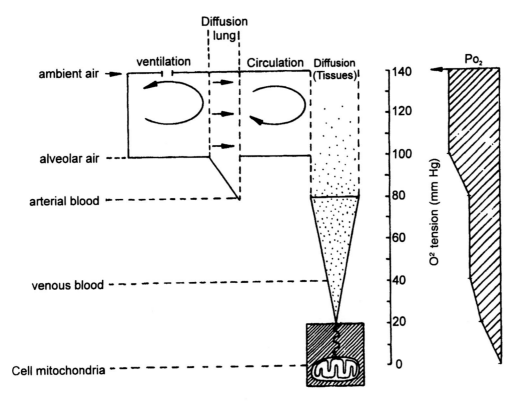

Fig. 3 Cascade flow of oxygen from a convectively ventilated lung to mitochondria in an air-breather. The ventilatory and circulatory systems maintain a partial pressure gradient across the air/blood interface, with the mitochondria serving as terminal sinks of molecular oxygen. The partial pressure of oxygen decreases from the lung, the tissue cells, and the mitochondria. The components of the pathway are structurally and functionally well sized to optimize the flow of oxygen (from Wood and Lenfant 1976).

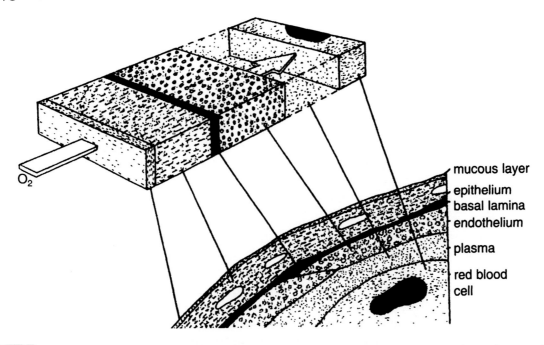

Fig. 4 Schematic view of the basic design of a gas exchanger. An external respiratory fluid medium is separated from an internal one by a thin composite tissue barrier. A partial pressure gradient that prevails between the two media drives oxygen from the external medium across the barrier to the blood. The accompanying drawing illustrates the structural components of the water-blood (tissue) barrier of the fish gills.

pressure gradient of oxygen prevails between the two compartments [Fig. 4]. The highly reduced form of a gas exchanger offers a useful conceptual and practical model for discerning the origin and subsequent changes that have occurred in the development and structure of respiratory organs. The extant prozoans with their plain cell membrane still display the most elementary functional design. Protoplasmic streaming, a normal circulation-like process in living cells, enhances the influx of oxygen.

After about 2 billion years of evolution of aerobic metabolism, the influx of oxygen and efflux of carbon dioxide still occur by passive diffusion at the tissue and cellular levels. Respiratory structures evolved from a common entity, apparently the plain, simple cell membrane. In the realization of optimal gas exchangers, some structural features and functional processes were conserved while others were overhauled. The congruities in the design of respiratory structures may indicate the importance of respiration for life: an immutable design is compelled. In a respiratory entity, there need only be a PO_2 between the gas-exchange media and an adequately thin tissue partitioning between them to allow diffusion of respiratory gases. Fundamentally, the respiratory medium used, the structural properties as well as the composition and the organization of the cellular and tissue elements are immaterial. Illustratively, such animals as rats, cats, and dogs have been kept alive breathing isotonic liquids enriched with oxygen (especially perfluorocarbons) at a pressure equivalent to that of air at sea level. Survival times have depended on such factors as experimental temperature and chemical composition of the fluid used. Death does not result from anoxia but rather from limitation of eliminating carbon dioxide. In a dog breathing hyperbarically oxygenated saline, arterial partial pressure of carbon dioxide (PCO_2) of 9.3 kPa (kiloPascals) and a pH of 7.2 was measured. Mechanical ventilation of the lung and addition of tris(hydroxymethyl) amino methane (a compound that minimizes the harmful effects of carbon dioxide accumulation) to the experimental solution reduced carbon dioxide loading, averting acidosis during liquid-breathing. With an arterial oxygenation of 32 kPa, e.g. in a dog, evidently oxygen delivery to the tissues during liquid-breathing is more than adequate.

On average, the vertebrate lung is estimated to utilize as much as 10% of the total body oxygen consumption. By default, the respiratory organs are the only structures in an animal's body in which a real "conflict of interest" between oxygen consumption and transmission to the tissue cells can occur. For optimal function, the design of a respiratory organ must allow the transfer of oxygen with minimal utilization by the organ itself. To achieve that, the least quantity of tissue must be committed in the construction of a gas exchanger. This must occur without the structural and functional integrity of the organ being compromised. The blood-gas (tissue) barrier in the avian lung [Chapter 8] and the thinnest (respiratory) parts of the tissue barrier of the interalveolar septum of the mammalian lung [Chapter 9] consist of an epithelium, a common basement membrane, and an endothelium. An interstitial space with smooth muscle and connective tissue elements such as collagen and elastic tissue occurs in the amphibian and the reptilian lungs as well as in the thick (supportive) parts of the tissue barrier of the mammalian lung.

Though best known for their gas exchange role, gills and lungs are multifunctional organs. In certain cases the nonrespiratory functions are as important as the respiratory. For example, gills are involved in osmoregulation, ammonia and urea excretion as well as regulation of levels of some blood-borne chemical factors such as hormones. The mammalian lung is an important site for synthesis, metabolism, and regulation of concentrations of various pharmacologically active agents that include prostaglandins, serotonin, bradykinin, angiotensin, and lipids such as the pulmonary surfactant. Oxygen consumption of the vertebrate lung itself can be attributed to the metabolic processes. While the design for gas exchange obligates minimal tissue infrastructure so as to particularly generate a thin water/blood-gas barrier, the nonrespiratory functions require a critical tissue mass to perform. For optimum performance, the ultimate design of a respiratory organ must reflect the various (in certain cases) conflicting demands placed on it. Methodical trial-and-error, cost-benefit analysis, trade-offs, and compromises are transacted so as to overcome or accommodate inherent limitations. Compared with other organs and organ-systems, respiratory structures are unique in that no cells or structural elements are ubiquitous to them as, for example, the hepatocytes are to the liver, the osteocytes to bone, or the neurones to nervous tissue. As a general rule, a respiratory organ can only be sufficiently defined from the perspective of its function (i.e., gas exchange) rather than from its structure. Elegant experiments have demonstrated the otherwise morphologically obscure respiratory sites (areas of high oxygen influx) in which suitable flagellates, e.g. *Polytoma*, were used as oxygen biomarkers. The catholic nature of the structure of gas exchangers may be ascribed to the fact that simple passive diffusion of respiratory gases should not prescribe an invariant design: an adequately high partial pressure gradient is basically all that is necessary to drive oxygen. For example, in a process that may be termed "reverse diffusion," air-breathing fish which can achieve a high level of arterialization of blood using accessory respiratory organs, succumb to asphyxia when they move into hypoxic or anoxic waters by losing oxygen from their blood and tissues, especially across the gills, into the surrounding water!

It is important to recognize that the modern respiratory structures are outcomes of a long evolutionary process. Through the evolutionary continuum, various designs have been fashioned and refashioned while being rigorously tried and tested. Unsatisfactory features and aspects have been discarded and optimal ones conserved. The various ultimate designs have been set by the need for oxygen to support particular metabolic capacities and life styles. In biology, the principles of homology and analogy are fundamental to understanding the mechanism of adaptation as set by natural selection. In the process of optimizing design and function, superfluous structures have undergone irrevocable deconstruction, with some becoming vestigial and others being totally eliminated.

The principal factors that have decreed the designs and constructions of respiratory organs are: a) habitat occupied, b) respiratory medium utilized, c) phylogenetic level of development, and d) lifestyle pursued. Through appropriate sizing, geometric patterning, and arrangement of the constitutive structural components, the designs have been continuously shaped and refined for efficient acquisition of oxygen. Basically, the cost-effective designs are those that have in financial language, required the "least capital outlay". In biological terms, such designs will entail mimimal cost to develop, operate, and maintain. The advent of energetic animals, particularly the endothermic-homeotherms (mammals and birds) should have compelled parallel refinement of respiratory organs. In amphibians [Chapter 6], remarkable transformations in terms of form, location, and function of the respiratory organs occur during metamorphosis. The external gills and the skin are the sole respiratory organs during the larval stages of aquatic development while on land, lungs replace them. Fish that live in hypoxic water have acquired the capacity for extracting oxygen from air and water. The bimodally breathing fish [Chapter 5] essentially live at the air-water interface, deriving benefits of aquatic and terrestrial existence.

The adaptive changes in the design of the respiratory organs and gas-exchange strategies have resulted by either: a) careful remodeling of default structures, b) development of auxiliary respiratory structure(s), or c) commissioning of an existing nonrespiratory structure to a respiratory role. For animals moving from water to air, due to the great physicochemical differences between the two media, the first option was difficult, if not impossible to follow: a transitional stage wherein water- and air-breathing organs could coexist was imperative. Most air-breathing fish have utilized the second strategy [Chapter 4]: structures such as the buccal cavity, gastrointestinal system, labyrinthine organs, and suprabranchial chamber membrane serve as accessory respiratory organs. The last strategy is exemplified by the swim bladder, which in some fish has been transformed to an efficient respiratory organ. Structural remodeling of respiratory organs creates new polarities, granting new functional capacities. Such specializations offer new ecological opportunities. Regarding respiratory adaptation, as a general rule, only one pathway or structure predominates at any one time.

In the obligate air-breathing fish, for example the lungfish *Lepidosiren paradoxa* and *Protopterus aethiopicus*, gill development has been greatly reduced. The remaining structures are vestigial and only serve the role of eliminating carbon dioxide.

Respiratory organs possess the following shared structural features: a) internal subdivision and/or stratified arrangement—means by which an extensive surface is generated, b) marked vascularization of the surface, and c) thin partitioning between the respiratory media, water/air and blood. Water (a liquid over the biological range of temperature and pressure) and air (a gas under similar conditions) are the only two naturally occurring respirable fluid media. The modern respiratory organs have evolved to utilize one or the other and in rare cases both. Compared with air-breathing, water-breathing is the primordial mode of respiration. As a respiratory medium, air is more amiable: it is 50 times less viscous than water; the concentration of dissolved oxygen is about 30 times that in water; the rate of diffusion of oxygen is 8×10^3 times greater than in water; and the capacitance coefficient, i.e., rise in concentration per increase in PO_2, is 30 times higher in air than in water. In saturated water, e.g. at 20°C, 1 ml oxygen is contained in 200 g water while 1 cm^3 oxygen is present in 5 ml of air (mass, 7 grams).

Lungs are thought to have evolved as adaptations to hypoxic or anoxic conditions in the early hydrosphere. The physicochemical differences between water and air have so differently shaped the morphologies of water-breathing (gills) and air-breathing (lungs) organs that usually those structures efficient in one fluid medium fail dismally in the other. The phrase "like a fish out of water," used to describe the perils an individual faces outside of its familiar habitat, depicts how a fish soon succumbs to asphyxia (lack of oxygen) when removed from water. The high viscosity, low concentration, and slow diffusivity of oxygen in water curtail effective breathing of liquids using lungs. During laminar flow, a liquid breather has to expend 60 times the energy required for breathing air. The maximum expiratory flow of a liquid is 40 to 100 times lower than air ventilation. When breathing liquid substances, mice, rats, dogs, and cats die from exhaustion of the respiratory muscles and subsequent accumulation of carbon dioxide to toxic levels.

4 EVAGINATION AND INVAGINATION: PROGNOSTIC DESIGNS OF THE RESPIRATORY ORGANS

In terrestrial vertebrates, respiration is confined to particular parts of the body. A conspicuous respiratory system cannot be clearly delineated below the level of mollusks and arthropods. Basically, respiratory structures have either developed as evaginations (= outfoldings = outpocketings) away from the surface or invaginations (= intuckings = cavitations = sacculations) into the substance of the body [Fig. 5]. The gills, the inaugural respiratory organs, fall in the first category and the lungs in the second. For efficient water conservation on land, invagination of the respiratory organs was imperative. By minimizing or eliminating the threat of desiccation, this allowed transition from an aquatic to a terrestrial habitat. Although circumstances and conditions have since changed, the risk of desiccation on terra firma is profound. For example, if the presently, human lungs were evaginated (i.e., protruded outwards like the external gills of fish) and the respiratory surface were spread out and directly exposed to the atmosphere, even in a moderately desiccating environment the water loss would be about 500 $L^{-1}day^{-1}$. This is some 1,000 times more than the normal loss in the invaginated organ. Under such conditions, a person would die of desiccation within about 3 minutes! Through intense internal subdivision of invaginated respiratory organs, an extensive surface area is created in a limited space. For example, while a sphere of a volume of 1 cm^3 has a surface area of 4.8 cm^2, 1 cm^3 of the parenchyma of the lung of the shrew *Sorex minutus* (a very small, highly metabolically active mammal) has an alveolar surface area of 2,100 cm^2. This is a factorial difference of 440. In the human lung, a respiratory surface area of about 140 m^2, often a bit overexaggerated (aptly for ease of

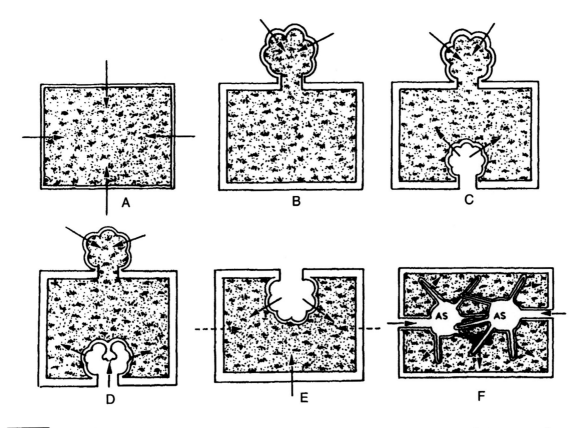

Fig. 5 Fundamental designs of gas exchangers. The unicells (A) have a plain gas exchanger in which diffusion of oxygen occurs entirely across the cell membrane. In more advanced animals, gas exchangers have formed either as evaginations (B) generally called gills and specialized for water-breathing, or invaginations (Figs. C-F) generally called lungs that are designed for air-breathing. (B), example of a unimodal water-breather; (C), bimodal breather with "unspecialized" accessory respiratory organ; (D), bimodal-breather with "modified" accessory respiratory organ; (E), terrestrial air-breather with the lung as exclusive gas exchanger (in amphibians diffusion occurs across the skin—shown by arrows); and (F), insectan tracheal system in which air is delivered directly to the tissue cells in large or highly metabolically active species across air sacs (AS).

conceptualization) to be equivalent to that of a tennis court, is packed in only about 4.5 L of the volume of the lung and in turn that of the thoracic cavity. The intensity of internal partitioning and hence the mass-specific respiratory surface area, quantitatively expressed as the surface density of the blood-gas barrier, correlates positively with the metabolic rates of vertebrates.

Certain functional advantages and limitations accompanied the development of internalized respiratory organs from externalized ones. As mentioned above, water conservation was the main advantage derived from invagination of respiratory organs. While evaginated respiratory organs (gills) can be ventilated unidirectionally and continuously produce a highly efficacious countercurrent system, lungs (invaginated = cul-de-sac = dead-ended respiratory organs) can only be periodically (= bidirectionally = tidally = in-and-out) ventilated [Figs. 6, 7]. Internalized respiratory organs cannot fully exploit the high ambient PO_2 since the inspired air is greatly diluted by the stale residual (dead space) air in the upper airways. The driving pressure drops from 21 kPa in the atmosphere to about 13 kPa at the respiratory surface. This translates into a loss of about one-third of the initial head (driving) pressure. In a resting person, where the dead space is about 140 cm³, about 28% of the 500 cm³ of the inhaled air does not reach the alveolar level. On the plus side, however, tidal ventilation can allow creation of isolated respiratory micromilieus. In the lung, the

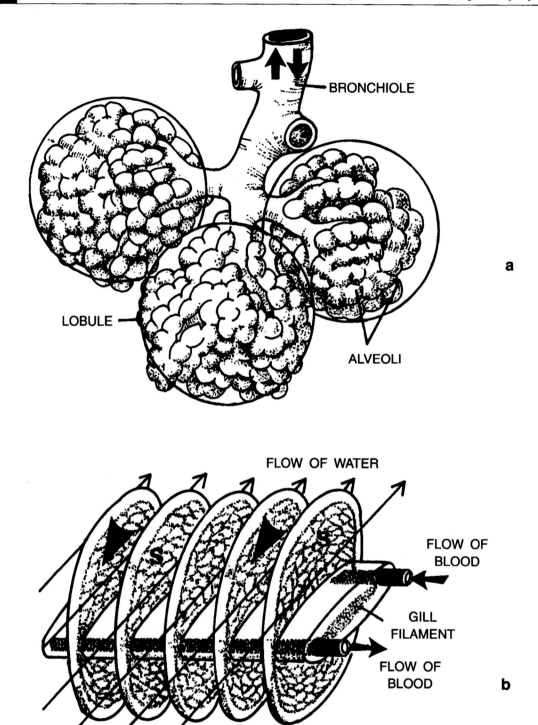

Fig. 6   Flow of respiratory media in an invaginated gas exchanger (a) and an evaginated one (b). Invaginated organs (lungs) are tidally ventilated while evaginated ones (gills) are unidirectionally ventilated; secondary lamellae(s). In gills, the direction of the blood flow in the secondary lamellae (➤) runs counter to that of the water in the interlamellar space (arrows) (Modified after Kylstra 1968).

concentration of carbon dioxide in the terminal respiratory units is higher than in the atmospheric air. The high concentration of carbon dioxide in the vertebrate lung is utilized in the bicarbonate (HCO_3^-) ion mediated buffer system for pH regulation. Production of confined microenvironments is not feasible in evaginated respiratory organs.

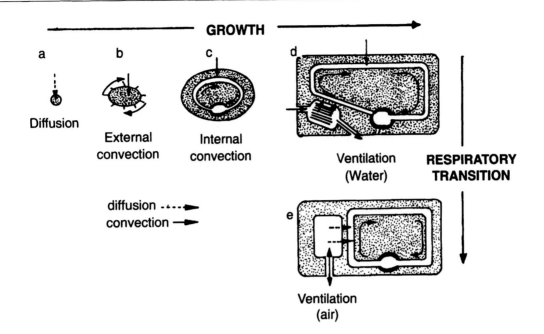

Fig. 7 Progressive development of the respiratory processes from unicellular to multicellular organisms. Diffusion is the only means by which oxygen is procured in unicells (a). External (b) and internal (c) convective processes are added on as animals increase in structural complexity. In gills, water and blood flow in opposite directions, constituting a countercurrent arrangement (d) while lungs are tidally ventilated (e) (after Burggren and Pinder 1991).

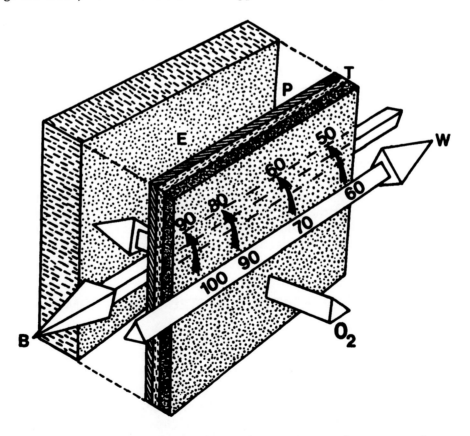

Fig. 8 Schematic diagram showing a countercurrent arrangement between respiratory media that contain oxygen at different partial pressures flowing in opposite directions. A diffusional gradient occurs over the length and duration of exposure, producing a highly efficient respiratory design. A countercurrent arrangement occurs in the fish gills. Curved arrows, oxygen flowing down a concentration gradient; T, tissue barrier; W, water; E, erythrocyte; P, plasma; B, blood.

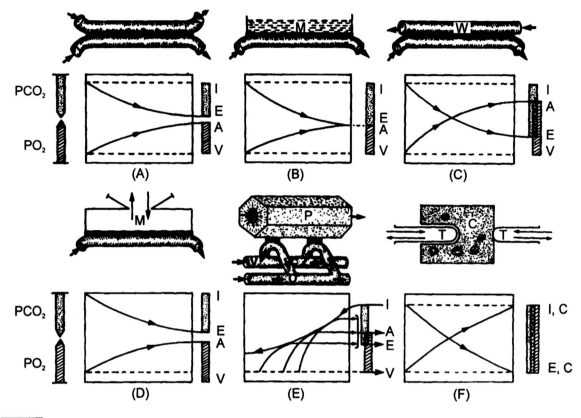

Fig. 9 Schematic illustrations showing the exposure of respiratory media in different gas exchangers. Respiratory efficiencies are shown by PO_2 and PCO_2 profiles in the inspired air (I), expired air (E), venous blood (V), and arterial blood (A). (A): concurrent system—gas exchange media flow in the same direction and oxygen uptake is very poor; (B): the skin—gas exchanger perfused but not actively ventilated; (C): counter-current arrangement—PO_2 in the arterial blood exceeds that in the end-expired air; (D): uniform-pool—influx of oxygen depends on prevailing ventilation and perfusion inequalities; (E): crosscurrent system in the avian lung— through the multicapillary serial arterialization system, the PO_2 in the arterial blood may exceed that in the end-expired air; (F): in the insectan tracheal system, oxygen is delivered directly to the tissue cells (by the tracheoles (T) with the PO_2 at the cellular level being essentially equal to the ambient. The PO_2 in the arterial blood may exceed that in the expired respiratory medium only in models C and E. m, respiratory medium which, depending on the type of animal, could be air or water: p, parabronchus; c, cell; v, blood (venous); w, water. In the schematic drawings, single arrows show directions of flow of respiratory media and diffusion of oxygen; arrows running in opposite directions show the gas exchangers that are tidally ventilated.

5 GEOMETRY AND ASSOCIATION OF STRUCTURAL COMPONENTS IN THE RESPIRATORY ORGANS

In practical terms, given a certain quantity of structural materials, a structural engineer can design and fabricate infinitely many devices. However, if parameters such as size, cost, strength, and safety factor of operation are defined, the appropriate form is rendered more precise. Moreover, the soundness of the contrived structure not only depends on the measure and the nature of the materials used in the construction, but also on the manner in which the individual parts are arranged and connected together. These fundamental principles of human design and construction can be equally applied to the fabrication of biological tissues, organs, and organ-systems.

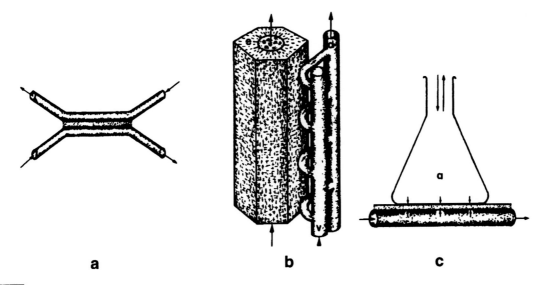

a   b   c

Fig. 10 Schematic drawing showing relationships between air and blood flow in three gas exchangers: a, fish gills; b, avian lung; c, mammalian lung. In fish gills, the air and blood flow in counterdirections. In birds the flow between the parabronchial venous blood and the airflow in the parabronchial lumen is crosscurrent. In the mammalian lung, the flow is described as ventilated pool. e, parabronchial gas exchange tissue and airflow in the lumen, ↑; v, venous blood flowing into the exchange tissue, ↑; a, (in Fig. b) arterial blood; In Fig. c: a, alveolar air; ↓, diffusion of oxygen across the blood-gas barrier: →, direction of alveolar capillary blood flow; parallel arrows, tidal ventilation of the mammalian lung. The efficiency of gas exchangers depends to a great extent, on the presentations between the gas exchange media.

In respiratory organs, the geometric association of the structural components determines the movements and presentations of the gas-exchange fluid media within and across the organs. If the respiratory media, i.e., the external (ventilating medium—air/water) and the internal (i.e., perfusing medium—blood) flow in the same direction, the arrangement is termed concurrent; if they run in opposite directions, it is called counter-current [Figs. 6, 8-10]. The scheme is designated crosscurrent if the fluid media flow in direction perpendicular to each other [Figs. 9, 10]. When the external medium of which the PO_2 is fairly uniform fronts a respiratory surface or organ, the configuration is termed a uniform pool [Figs. 6, 9, 10]. The qualitative and quantitative attributes of the structural components in different respiratory organs bespeak the efficiencies and limitations inherent in the various designs. The remarkable efficiency of the countercurrent system of fish gills was necessary for survival in water, a fluid medium deficient in oxygen and one which compared to air (due to its greater viscosity) is more costly to breathe. The oxygen extraction ratio in fish gills is as high as 92%. If direction of water flow is experimentally reversed, i.e., a countercurrent system is converted to a con-current one, oxygen extraction falls to below 10%. In the avian lung [Chapter 8], the crosscurrent arrangement between the pulmonary capillary blood and the flow of air in the parabronchial lumen [Figs. 8, 9] provides the most efficient respiratory organ among the air-breathing vertebrates. Through a serial multicapillary additive arterialization arrangement [Figs. 8, 9], in some conditions (e.g. hypoxia and exercise), the PO_2 in the arterial blood may exceed that in the end-expired air. Likewise, the PCO_2 in the end-expired air may exceed that in the end-capillary blood. Of all the evolved gas exchangers, regarding PO_2, such overlap is only possible in the countercurrent system of fish gills.

6 FRACTAL GEOMETRY: QUINTESSENTIAL DESIGN IN NATURE AND IN THE RESPIRATORY ORGANS

In biology, those conserved features and processes that diffuse across taxa are normally the most consequential ones for survival. With the primary role of the respiratory organs being that of appropriating molecular oxygen, certain encompassing features occur in them. A large surface area is achieved by subdivision of the air-conducting and blood-conducting elements [Plates I, II]. Fractal geometry is a powerful method of defining and quantifying the form and shape of complex biological structures. Entities with fractal properties include coastlines, viruses, tissues, and cells: they lack absolute spatial dimensions.

Classical (i.e., Euclidean) geometry defines space in terms of discrete linear (i.e., integer) dimensions: a point has no (zero) dimension, a line has one dimension, a plane (surface) has two, and a solid (volume) has three dimensions. In a 3-dimensional space, topologically a folded structure is in progressive transition between a smooth, flat surface (i.e., a 2-dimensional form) and a volume (i.e., a 3-dimensional form). A contorted structure should have a fractal dimension between 2 and 3. Iterating algorithms inscribed in the genome regulate development of "space-filling" structures such as respiratory and glandular organs. Morphogenetically, fractal programming minimizes the possibility of error during development. To sufficiently delineate the topological attributes of morphologies with fractal attributes, fractal (fractional) power dimensions are necessary. Propounded by Benoit Mandelbrot in 1977, fractal geometry provides a powerful tool for investigating, interpreting, and understanding form. Respiratory organs are fundamentally constructed by fashioning together three structurally fractal entities, namely the arterial, venous, and bronchoalveolar systems [Plates I-IV]: they manifest regular dichotomous branching. By intimately interweaving with each other, an extensive surface area is generated and the respiratory media are closely exposed to each other. Fractal design allows biological structures to tolerate various perturbations without failure. Such malleability confers greater error tolerance, imparting high safety margins of operation. The structural and functional plasticity inherent in the fractal design makes it easier for organisms to engage selective pressures and institute appropriate adaptive measures. It allows for a more or less trial-and-error evolutionary process.

Though the lung appears to be rather passive, especially when compared to the closely located mechanically active heart, intrinsically it (lung) is not an inert organ. It is metabolically highly active; it provides an interface between blood and air, fluid media that are biophysically remarkably different; it is the only organ in the body that transmits the entire volume of blood in the systemic circulation; it is subjected to changing hemodynamic pressures; and it submits to ventilatory rhythms under movements of the ribs and diaphragm. The surface fractal value of the mammalian lung is about 2.5, with that of the bronchial tree about 3. In the human lung, the pulmonary vascular system is so greatly folded that it has an effective fractal dimension of 3, with the arteries alone having that of 2.7. The fractal dimensions of the diameter-element of the arterial and venous trees in the human lung are respectively 2.71 and 2.64 while equivalent values for the length-element are 2.97 and 2.86 respectively. In the dog lung, the fractal dimension of the blood flow is 1.22. The high fractal values of the airway and vascular elements of the mammalian lung confirm that the structures are greatly folded and branched [Plates I-IV]: they virtually fill a 3-dimensional space, i.e., the volume of the lung and in turn that of the thorax. The fractal nature of the design of the lung permits a large internal surface area to be homogeneously ventilated with air and perfused with blood at low energy cost. Moreover, the property may allow the pulmonary vasculature to be highly distensible, permitting it to contend with changing blood pressures. Pulmonary blood flow is pulsatile from the entrance of the pulmonary circulation to its outlet in the left atrium. Although only about 9% of the total blood volume is contained in the heart, the distensibility of the pulmonary circulation is of particular functional significance since it permits transient imbalance between the outputs of the left and right ventricular chambers of the heart. A high fractal dimension of about 2.5 in the lung-air sac system of birds [Chapter 8] may partly explain how certain birds are able to fly at high altitudes without failure and without having had to institute exceptional modifications in the structure of their respiratory organs. For such a highly diversified taxon (about 9,000 species), except for differences in size and location of the air sacs, pneumatization of the long bones, and development of the parabronchial system, the

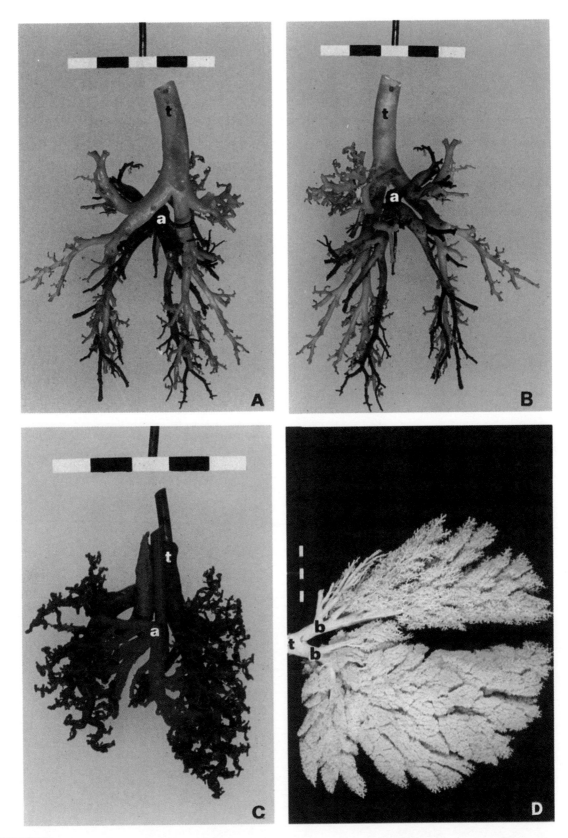

Plate I A, B: Photographs of anterior and posterior views of preparations made by injection with plastic fluid (setting) material (a cast) of the tracheobronchial (t) and arterial (a) systems of a rabbit lung showing the congruity in the morphological patterning of two conducting systems, namely the airway and arterial systems. C: Posterior view of the arterial (a) and tracheobronchial (t) systems of a rat lung. D: Anterior view of a cast of the tracheobronchial system of the lung of a dog showing its fractal scheme. t, trachea; b, principal bronchi. Scale bars, 1 cm.

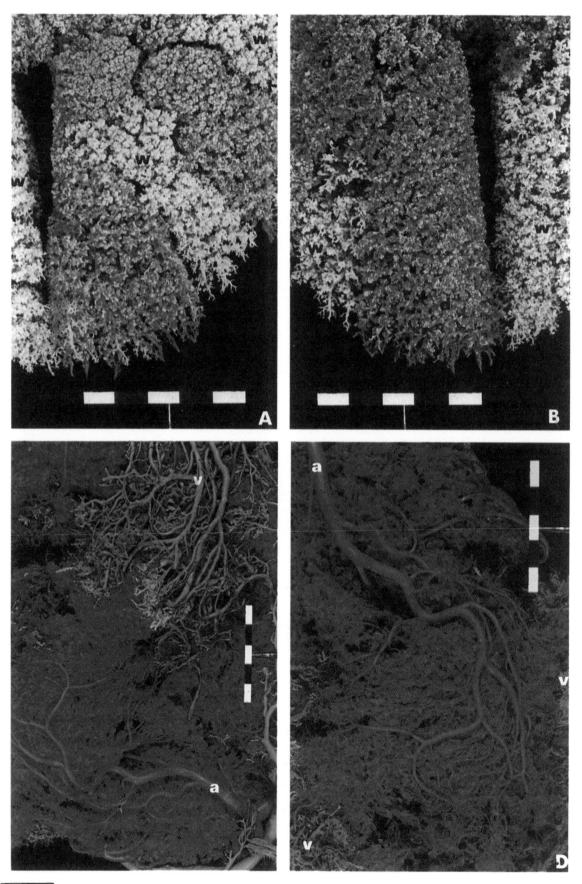

Plate II A, B: Photographs of preparations of a rabbit lung mode by injection with plastic fluid (setting) material showing the close relationship between the airway and arterial sytems of the lung [white areas (w): only the airways were cast; red areas (d): both airway and arterial systems were cast]. C, D: Fractal scheme of the arterial (red) and venous (blue) systems of the heart of the African elephant, *Loxodonta africana.* Scale bars, 1 cm.

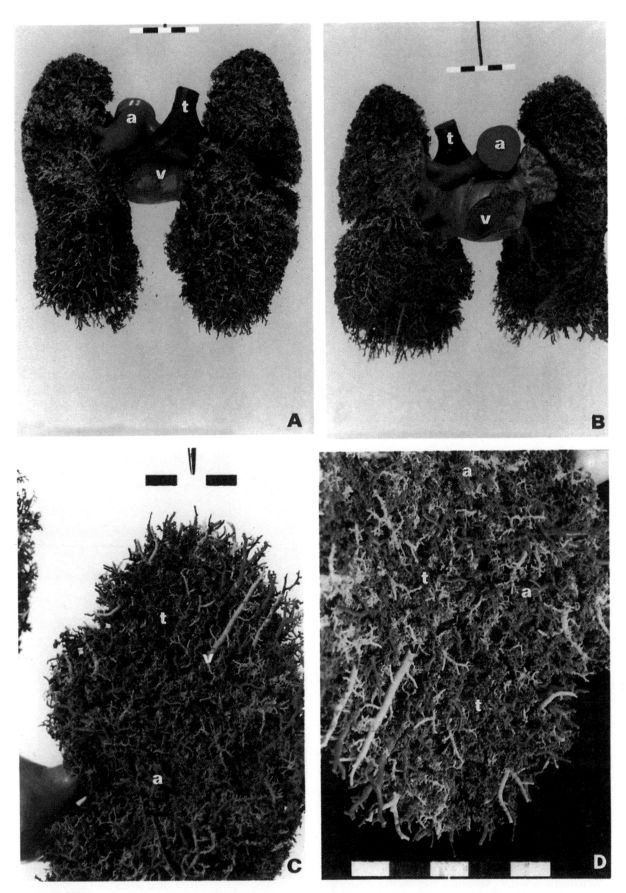

Plate III A, B: Photographs of preparations of human lung made by injection with plastic fluid (setting) material showing the arterial (a), tracheobronchial (t), and venous (v) systems. The structures interact very closely and display fractal designs. A: Anterior view; B: Posterior view. C: Superior lobe of right lung. D: Close-up of lower edge of inferior lobe of left lung. Scale bars, 1 cm.

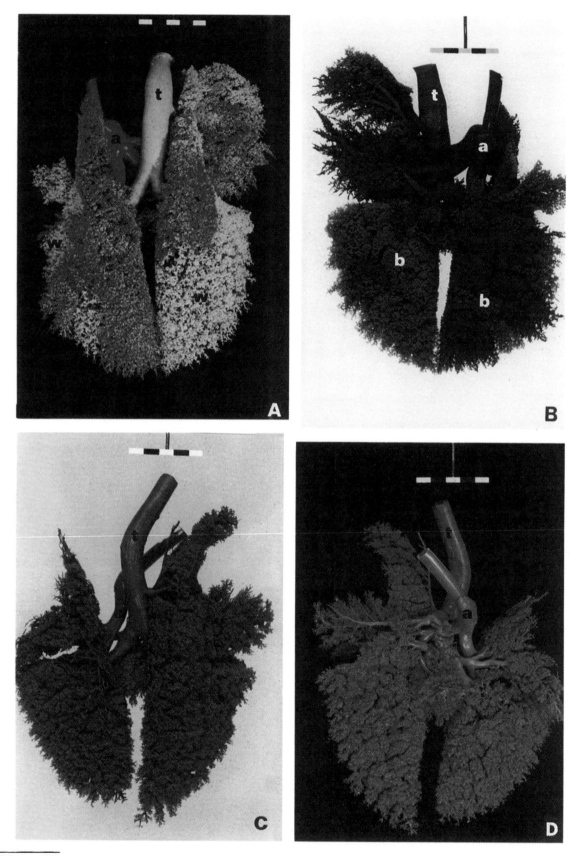

Plate IV A, B, C, D: Photographs of lung of a dog prepared by injection with plastic fluid (setting) material showing the subdivisions of the tracheobronchial (t) and arterial (a) systems. A: Anterior view; B: Posterior view; C: Anterior view; D: Posterior view. A: white areas (w) show regions where only airways were cast and red areas (r) those areas where both airways and arteries were cast. B: blue areas (b) show regions where only airways were cast. In C and D, where cast material of same color was used to cast the airway and arterial systems, it is not possible to differentiate the two due to their morphological congruity. Scale bars, 1 cm.

design of the avian respiratory system is remarkably uniform. In the secondary lamellae of the fish gill [Chapter 2], the assumed small fractal dimension of 2 is based on the view that their surface is smooth. Microridges characterize the secondary lamellae, however, the respiratory units of most fish gills [Chapter 2]. They even occur in the gills of the most ancient fish—the coelacanth *Latimeria chalumnae* and the sturgeon *Acipenser transmontanus*. Microridges are poorly developed, however, in the gills of such fish as *Trachurus mediterraneus* (a carangid fish) and lacking in the secondary lamellae of hill-stream fish such as *Danio dangila*. Like other vertebrate respiratory organs, the surface of the gills of most fish has a fractal construction. The apparently lower fractal dimensions in the lung-air sac system of birds and fish gills (compared with those of structures of the mammalian lung) may be explained by the fact that the two kinds of gas exchangers respectively have highly efficient crosscurrent and countercurrent systems inbuilt in them.

7 STRUCTURAL-FUNCTIONAL CORRELATIONS IN THE DESIGN OF RESPIRATORY ORGANS

Respiratory organs afford the first interface between oxygen in the ambient environment and metabolic machinery of the body. The design of the constitutive components should determine the quantity of oxygen that can be procured and utlized to generate energy. It is hence axiomatic that the aerobic capacities of an animal should be commensurate with the efficiency of the respiratory organ(s). Charged with a common function (gas exchange), respiratory organs manifest shared morphological and physiological features. Regarding structure, the ubiquitous ones are an extensive surface area, a thin water/blood (tissue) barrier partitioning between the respiratory media (water/air and blood), and intense vascularization. Functionally, through ventilation and perfusion, a partial pressure gradient of oxygen is maintained across the tissue barrier of a gas exchanger.

Form and function of respiratory organs are inextricably interrelated. Morphology is the outcome of diverse yet integrated functions. Small, metabolically active endotherms, such as shrews, small bats, and hummingbirds, have highly specialized respiratory organs. In mammals (a taxon better studied among vertebrates), the lung and skeletal muscles and their mitochondria are integratively "built" to meet the requirements for oxygen. The interdependence between structure and function for adept performance is termed "symmorphosis". It is defined as: "A state of structural design commensurate with functional needs resulting from regulated morphogenesis, whereby the formation of structural elements is regulated to satisfy but not exceed the requirements of the functional system." Though so termed and propounded only about 20 years ago, the concept of symmorphosis is not wholly new to biology: the great Greek philosopher Aristotle (332 B.C.) observed that "nature does nothing to no purpose".

In gills [Chapter 2], a large respiratory surface area is generated through a stratified design. In teleosts, for example, four pairs of gill arches give rise to hundreds of gill filaments that in turn produce thousands of secondary lamellae. In lungs, a large respiratory surface area is generated by internal subdivision of the airway and vascular elements [Plates I-IV]. In the human lung, e.g. where the airway system branches dichotomously approximately 23 times before terminating at the alveolar level, there are 2^{23} (8×10^6) end branches. A thin partitioning between the respiratory media is achieved by a thinning of the epithelial lining down the vascular and airway systems, bringing water/air and blood into close proximity. In fish gills, an elaborate primary epithelium covers the surface of the gill rakers, gill filaments, and interlamellar spaces while a squamous epithelium lines the secondary lamellae. The upper airways of the avian [Chapter 8] and mammalian [Chapter 9] lungs are lined with a thick, pseudostratified, columnar, ciliated, secretory epithelium. With progressive bifurcation of the bronchial tree, the elaborate epithelium gradually changes to simple columnar, simple cuboidal, and ultimately into a squamous one at the alveolar/air capillary levels. In the lung of an adult person, for example, there are about 300 million alveoli. The thickness of the blood-gas barrier (harmonic mean) is 0.65 μm.

Among vertebrates, compared with ectotherms (fish, amphibians, and reptiles), structurally and functionally the endotherms (birds and mammals) have more specialized and efficient respiratory organs. The respiratory surface area and hence intensity of internal subdivision of the lungs increases from amphibian [Chapter 6], reptilian [Chapter 7], mammalian [Chapter 9] to avian ones [Chapter 8]. The highest mass-specific respiratory surface areas in vertebrate lungs have been reported in a bat [epauletted fruit bat, *Epomophorus wahlbergi* (138 cm^2 g^{-1})] and two species of bird [violet-eared hummingbird, *Colibri coruscans* and African rock martin, *Hirundo fuligula* (87 cm^2 g^{-1})]. The thinnest partitioning between the respiratory media in the vertebrate gas exchangers, the water/air-blood barrier, occurs in the avian and mammalian lungs with the barriers being much thicker in the fish gills. The barrier is relatively thinner in amphibian and reptilian lungs. The blood-gas barrier in *H. fuligula* is 0.090 μm and that in *C. coruscans* 0.099 μm thick. In mammals, the bat *Phyllostomus hastatus* has the thinnest blood-gas barrier of 0.120 μm. In air-breathing vertebrates, the thickness of the blood-gas barrier appears to have been optimized. This is shown by the fact that although mammals span a collosal range of body mass from the minute 2.6 g Etruscan shrew *Suncus etruscus* to the almost 150-ton blue whale *Balaena mysticetus* (a factorial difference of about 60 × 10^6), the thickness of the blood-gas barrier in the lung of the shrew of 0.27 μm differs from that of 0.350 μm of the whale by only a factor of 1.3. In birds, the thickness of the blood-gas barrier in the 7.3 g violet-eared hummingbird *C. coruscans* is 0.099 μm while that of the ostrich *Struthio camelus* (150 kg) is about 0.56 μm: the body mass factorial difference is 2 × 10^4 while that of the thickness of the barrier is only about 6.

Adaptively necessary to increase the respiratory surface area, internal subdivision of the lung occurs at a cost. In compliant lungs, narrow terminal respiratory components compel more energy to dilate. Moreover, such units have a high propensity for collapsing from the high surface tensional forces that prevail at the air-water interface. Dipalmitoylphosphatidylcholine (surfactant) is a phospholipid material that lines the respiratory surface. It lowers surface tension, providing stability to the air spaces. Trade-offs and compromises between maximization of the respiratory surface area, reduction of ventilatory cost, and reduction of size of the terminal gas-exchange components are delicately transacted in founding an optimal respiratory design. The presence of surfactant on the surface of the air capillaries of the virtually rigid avian lung [Chapter 8] is rather perplexing. The surfactant is known to play other roles, however, that include prevention of transudation of blood plasma onto the respiratory surface and protection of the epithelial lining from the destructive oxidative effect of molecular oxygen.

2

Gills

1 EVOLUTION OF WATER-BREATHING AND DEVELOPMENT OF GILLS

Life indubitably first formed in water. It provided both a stable environment and a necessary physical support to the simple, delicate life forms. Given to the harmful effects of ultraviolet radiation, during the first 150 to 200 million years life was totally confined to water. Primitive anaerobic microorganisms flourished in water in excess of 500 million years before molecular oxygen appeared. The distribution of water on Earth spatially and temporarily dramatically changed, especially after the breakup of the single landmass of Pangea and subsequent drifts of the continents. Regarding quantity and character (through the hydrologic cycle), water has remained fairly constant for about the last 3,000 million years. It has provided a conducive environment where life has flourished. More than one-half of the living vertebrates have arisen directly from lineages that still inhabit water. Fossil evidence strongly suggests that the earliest tetrapod land vertebrates (labyrinthodont amphibians) evolved from lobe-finned fishes of the class Osteichthyans. Because they do not lay amniote eggs in which the embryos themselves produce protective enclosing membranes, such as the amnion, fishes and amphibians are known as anamniotes. Gills are the primordial respiratory organs. Pharyngeal gills are a longstanding prototypical morphological feature of the phylum Chordata. Members of the long extinct ostracoderms date back to at least 550 million years. Over the time that they have inhabited water, fish have adapted and thrived very well. Extant fish include some 30 species of lampreys, 21 species of hagfishes, 100 species of chondrichthyans and perhaps as many as 30,000 species of bony fish (osteichthyans).

As respiratory organs, gills have been highly refined for water-breathing. The highly efficient countercurrent design of gills [Figs. 6, 8-10]—in which water flow across the secondary lamellae (the respiratory units of the gills) [Figs. 11-15] and blood flow in the vascular channels meet in opposite directions—provides a highly efficient respiratory design. Such an effective organ was necessary for survival in a fluid medium which, compared to air, is more viscous and relatively poorer in oxygen content. Gills, gas permeable outgrowths (i.e., evaginations) from the body [Fig. 5], present an interface between two compartments, an external fluid medium (water) and an internal (extracellular) one (blood). Vertebrate gills are classified into external gills, i.e., those that dangle freely into the surrounding water, and internal. Internal gills are covered by various cutaneous modifications such as an opercular flap (e.g. in teleosts) or mesenchymal tissue mass (e.g. in elasmobranchs—sharks and rays).

External gills are rare in adult fish but play an important respiratory role in amphibian tadpoles and in water-breathing neotenic forms, e.g. *Necturus maculatus*. In the larval forms of elasmobranchs and some larvae of Chondrostei and Teleosti, external gills are long filaments that float in the albuminous fluid in the eggcase. True external gills occur in the larval forms of such fishes as Polypteridae and Dipnoi (lungfishes). They are delicate thread- or featherlike structures. External gills are mainly ventilated by forward movement or by placement of the body across a current of water. They are highly susceptible to physical damage and restrict motion. The vascular shunt (anastomotic) blood vessels that occur between the afferent (dorsal segment of the aortic arch) and efferent (ventral segment of the aortic arch) arteries bypassing the external gills as in *Rana temporaria*, *Bufo bufo*, and the urodele amphibian *Amblystoma tigranum*, are thought to sustain circulation when the external gills atrophy and the animal switches from water- to air-breathing.

In general, amphibians possess gills during their larval stages of development. Urodela (Caudata) and Apoda (Gymnophiona or caecilians) have external gills while Anura (Salentia) have internal ones. Except for the external gills of the neotenic urodeles such as *Necturus* and *Ambystoma* that persist throughout life, amphibian gills are ephemeral (disposable) respiratory organs. The internal gills of tadpoles of Anura display rows of branched lamellae supported by gill bars separated by four gill slits. The gill slits close up with the onset of pulmonary breathing except for some Urodela that have readapted to aquatic life in which slits sometimes remain open throughout life.

Development of external gills of larval amphibians is determined by the oxygen tension in water. In hypoxic water, the gills of *Rana temporaria* increase in size while in well-oxygenated water they regress. The external gills of larvae of those salamanders that inhabit well-oxygenated water are less prominent: the respiratory surface area is small and the water-blood barrier is relatively thicker than in those that live in hypoxic (oxygen deficient) waters. In fish living in water with a PO_2 of 10.7 kPa, the gills are much larger than those under a partial pressure of 10 kPa. The external gills of amphibians differ remarkably from those of fish: macroscopically they form arborescent organs and are not arranged in an orderly, stratified manner. The gills of elasmobranchs lack skeletal support: columns of mesenchymal tissue confer support.

In amphibians, the contribution of gills to the overall respiratory process differs within and among species. Unlike in certain highly adapted air-breathing fish, amphibian gills are never adequately well specialized for gas exchange in air. Moreover, even in water, at no stage in life do they constitute exclusive respiratory organs. Before the ventilatory muscles develop, the external gills of newly hatched larvae of the African lungfish *Protopterus aethiopicus* have cilia that play a trophic role and assist in movement of water across them. During postlarval development, in *Protopterus ampibius*, the fraction of the total volume of oxygen acquired through the external gills and the skin decreases with development of the lung. In the brooding male South American lungfish *Lepidosiren paradoxa*, tuftlike, highly vascularized structures are found on the pelvic fins that serve as gills. They are considered important in parental rearing of eggs and larval forms in burrows constructed for that purpose. The structures are presumed to transfer oxygen from the blood of the male fish to the immediate vicinity of the eggs and the developing larvae in the hypoxic Amazonian waters that the fish inhabits. On hatching, the young larvae respire through external gills that start to atrophy after 45 days of life. The larvae come to the surface to breath across the lung at about the time the pelvic fin gills of the male lungfish start to regress.

2 STRUCTURE OF FISH GILLS

For remarkably divergent taxa, the basic structure of gills in agnathan and gnathostomatous fishes is strikingly similar. The fragile and yet elegant structure of gills has long fascinated biologists. Gills combine simplicity of design with functional efficiency. Being the primeval respiratory structures, they should provide an important design model for the study of the various evolved gas exchangers. In the predominantly air-breathing fish, e.g. mudskippers and lungfishes, gills are generally poorly developed. The gills of lungfishes (Dipnoi) differ remarkably from those of other classes of fish: they do not form regular arrays of primary and secondary lamellae but look rather like the external arborescent gills of a tadpole. Moreover, pillar cells are lacking. Much as they are structurally similar to those of fish that have hemoglobin in the blood, the gills of the hemoglobinless Antarctic icefish, e.g. *Chaenocephalus aceratus*, *Chamsocecephalus esox*, and *Chaenichthys rugosus*, have fewer secondary lamellae. Their skin is very highly vascularized and should play an important role in gas exchange.

The most elaborate gills are the internal ones of the bony fish. Typically, a hard opercular cover protects four pairs of gill arches. In the teleost fish, gill arches support gill filaments that in turn give rise to secondary lamellae [Figs. 11-15]. It is noteworthy that uncharacteristically, in the gills of *Trachurus mediterraneus*, the secondary lamellae are further subdivided. Strictly speaking, the secondary lamellae are the respiratory units of the fish gills. They are thin, flattened, hemispherical flaps bilaterally placed on the gill filaments [Figs. 11-15]. An elaborate epithelium (the primary epithelium) covers the gill filament while a simpler one (the secondary epithelium) lines the secondary lamellae [Fig. 15]. Pavement, epithelial, basal, interstitial, and endothelial cells constitute the water-blood barrier. Though typically thick, in most fish the barrier may be as thin as 0.2 µm in some parts of the secondary lamellae of some teleosts. The respiratory surface area and thickness of the water-blood barrier of the gills correlate with the metabolic capacities of the fish and the physical characteristics of the environment they inhabit.

Pavement, chloride cells (mitochondria-rich cells or ionocytes), and mucous cells are found in the primary epithelium. The pavement cells [Fig. 16] are squamous cells with microridges on their outer surface. Stretching greatly, they regulate the exposure of chloride cells to water [Fig. 16]. The patterning of the microridges differs between various areas of gills and between gills of different species of fish [Fig. 17, 18]. Microridges are lacking from lamellar epithelia of pelagic fish such as bluefish (*Pomatomus saltatrix*), Atlantic mackerel (*Scomber scombrus*), Atlantic bonito (*Sarda sarda*), and the hill-stream fish *Danio dangila*. Generally, the ridges decrease in size and frequency from the gill arch to the gill filament to the secondary lamellae. Microridges have been associated with diverse roles, viz. trapping and holding mucus, providing structural integrity to the gill epithelium, generating turbulence, reducing drag forces at the water-gill interface, and increasing the respiratory surface area. Moreover, they are thought to allow extension of pavement cells without their undergoing stress failure. This may be important for ensuring integrity of highly dynamic cells that are exposed to a medium of which the ionic composition and osmotic pressures may temporarily and spatially shift greatly. The size and the shape of the microridges are influenced by factors such as electrolytes, salinity, hormones, and hydrodynamics of water flow over the gills. The presence of microridge-like structures on the nonrespiratory tracts of the accessory respiratory organs of air-breathing fish such as the climbing perch *Anabas testudineus*, the snake-head fish *Channa striata* and the catfish *Clarias mossambicus* indicates that the accessory respiratory organs develop from the branchial arches.

Secondary lamellae are divided into vascular channels by polygonal pillar cells [Fig. 15]. Their cytoplasmic arborizations line the channels [Figs. 15, 19]. The pillar cells contain abundant contractile microfilamental actomyosin elements. Together with maintaining the structural integrity of the secondary lamellae, the pillar cells are envisaged to play an important role in regulating lamellar perfusion. Shunts bypassing the secondary lamellae en route from the ventral to the dorsal aorta do not occur in the common water-breathing teleosts. In a single gill filament, the blood flow through the marginal channels, i.e., the tip of a secondary lamella [Figs. 20-22] is less than that across those at the base. In the gills of the mudskipper *Boleophthalmus boddarti*, the water-blood barrier is thinner around the

Fig. 11   A scanning micrograph of the gill filaments of the teleost (rayfin fish) *Alcolapia grahami* showing gill filaments (f) and secondary lamellae (s). ➤, interlamellar spaces. Scale bar, 0.44 mm.

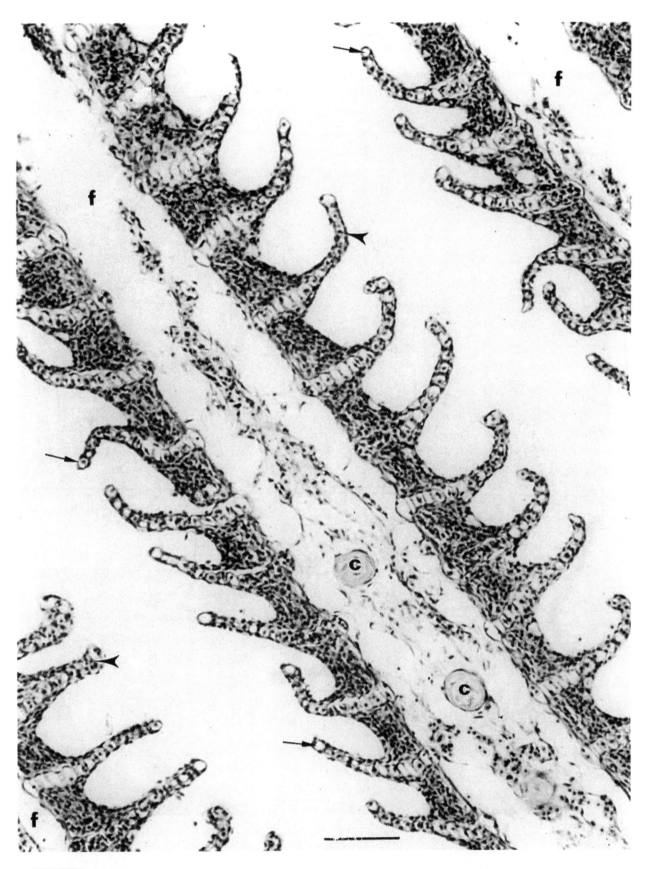

Fig. 12 A light microscope picture of the gills of a bimodal-breathing fish, *Clarias mossambicus*, showing gill filaments (f), secondary lamellae (arrowheads), and vascular channels (arrows) in the secondary lamellae. c, cartilages that support the gill filaments. Scale bar, 0.5 mm.

Fig. 13 Scanning electron microgroph of gill filament (f) of the teleost *Oreochromis niloticus* showing secondary lamellae (s); arrows, interlamellar spaces; arrowheads, intercellular junctions of the epithelial cells covering a gill filament. Scale bar, 0.3 mm.

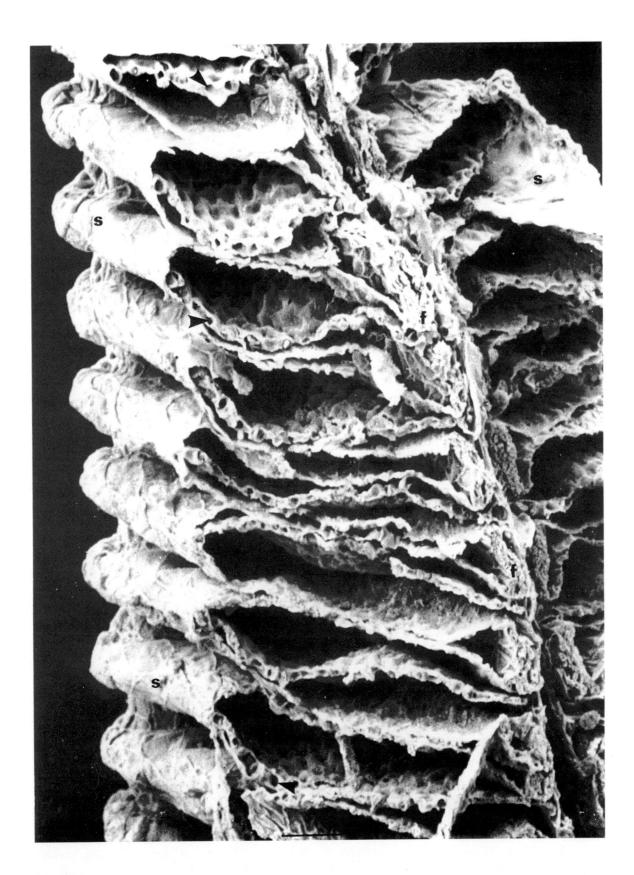

Fig. 14 Scaning electron microgroph of gill filament (f) of the teleost *Alcolapia grahami* giving rise to secondary lamellae (s). Arrowheads, vascular channels. Scale bar, 0.3 mm.

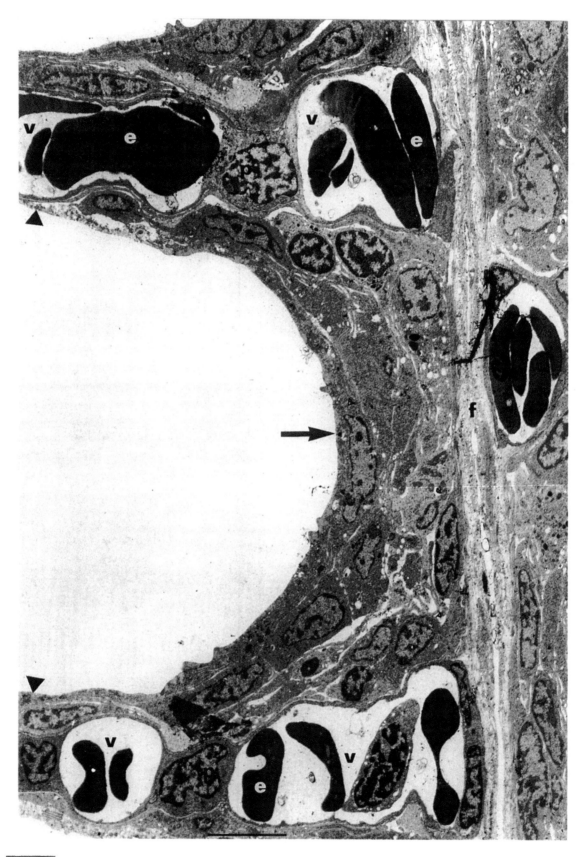

Fig. 15 Transmission electron micrograph of gill filament (f) of the teleost *Alcolapia grahami* giving rise to secondary lamellae (arrowheads). A composite epithelium is seen in the interlamellar space (arrow) while a thinner (arrowheads) one overlies the secondary lamellae. p, pillar cells; v, vascular channels; e, erythtocytes. Scale bar, 20 μm.

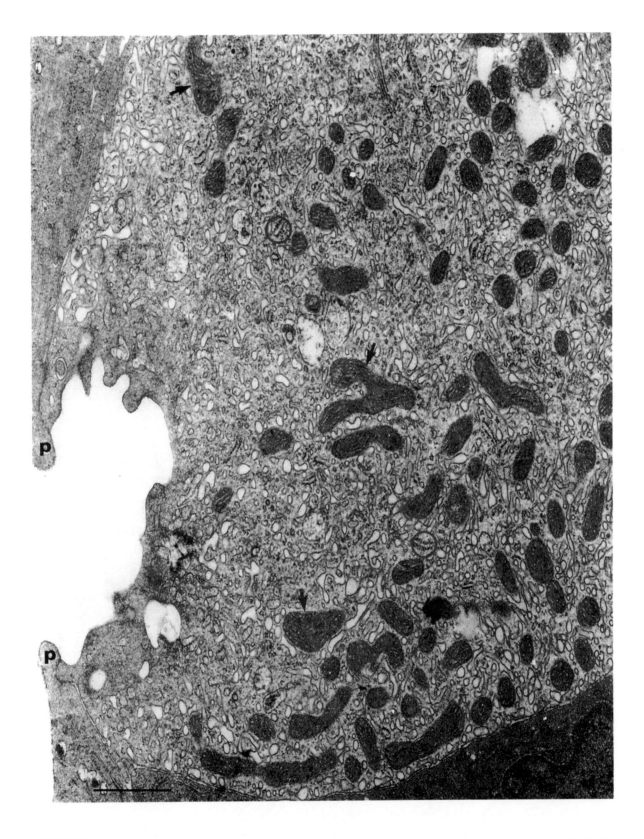

Fig. 16 Transmission electron micrograph of chloride cell of the gills of the teleost *Alcolapia grahami*. The cell has numerous mitochondria (arrows) interspersed between a dense microtubular network. p, pavement cells; s, supporting cells; circles, rough endoplasmic reticulum. Scale bar, 1 μm.

Fig. 17 Scanning electron micrograph of microridges (arrows) on a gill filament (f) and secondary lamellae (s) of the gills of the teleost *Alcolapia grahami*. Scale bar, 0.3 μm.

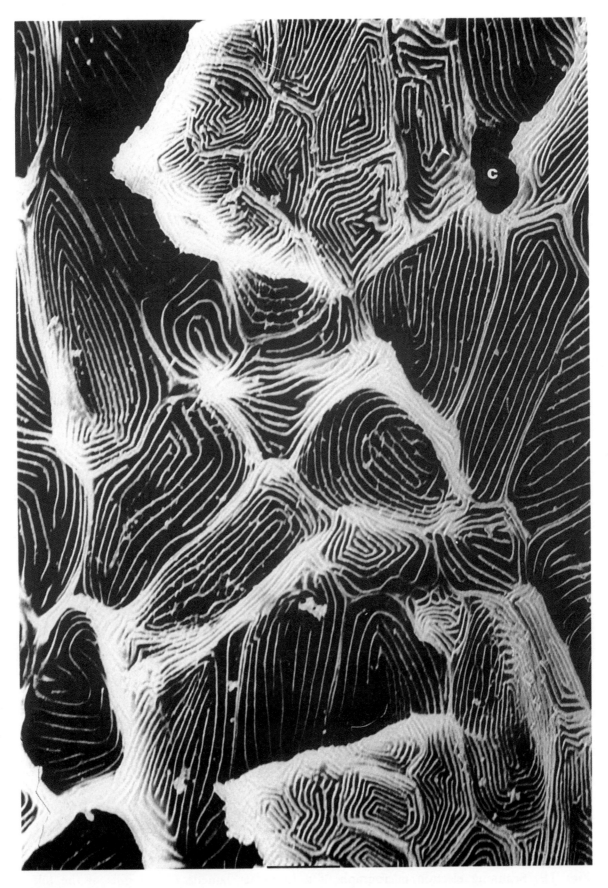

Fig. 18 Scanning electron micrograph of microridges on the external surface of a gill filament of the teleost *Alcolapia grahami*. c, pore of a chloride cell. Scale bar, 1 mm.

Fig. 19 Scanning electron micrograph of a secondary lamella of the gills of the teleost *Alcolapia grahami* showing vascular channels (v) that are separated by pillar cells (arrows). e, erythrocytes. Scale bar, 18 μm.

marginal channels that are better exposed to water. Certain pharmacological agents such as serotonin, adrenaline, and noradrenaline affect the flow of blood across the secondary lamellae. In the gills of the European eel *Anguilla anguilla*, the main sites of gill vascular resistance are at the level of the afferent lamellar arterioles and the secondary lamellae themselves. During its transit across the secondary lamellae, on average the blood is in contact with water for only about 0.5 seconds, a fleetingly short time but one that is more than adequate for complete oxygenation of the hemoglobin.

The mucous covering of the gills has been associated with various functions that include protection from mechanical damage, invasion by pathogens, and absorption and expropriation of toxic heavy metal ions. Other roles of the mucous covering are regulation of oxygen and electrolyte transfer across the gill epithelium. From their morphological features and topographical locations in the primary epithelium, two kinds of chloride cells occur in the gills of many teleosts. Dark and light as well as superficial and deep chloride [Fig. 23]. The dark chloride are generally located closer to the free epithelial surface of the gills while the light ones are found nearer the vascular channels. Both types of cells bear numerous mitochondria that are intercalated in the interstices of a dense microtubular network [Figs. 16, 24, 25]. In addition, an accessory chloride cell has been described in the gills of marine fish. The functional significance of the morphologically different chloride cells is presently not known. While it is quite possible that they may be one and the same cells at different stages of development, they could as well be functionally distinct, playing different roles in ionic exchange. The epithelial cells of the gills, especially the chloride cells, are highly sensitive to ionic and temperature changes in water. Thus the transfer of euryhaline fish from fresh water to sea water causes conspicuous but reversible changes in the chloride cell morphology, location, and numerical density. Hypercapnic acidosis causes an increase in number as well as apical surface area of the chloride cells while injection of cortisol increases the number of chloride cells by a factor of three.

3 FUNCTIONAL DESIGN OF FISH GILLS

Although best known for their respiratory function, fish gills perform diverse, seemingly unrelated functions. These include osmoregulation, acid-base balance, ammonia excretion, regulation of circulating hormones such as catecholamines and angiotensin, as well as detoxification of plasma-borne harmful substances. The structure of the gills must hence present a compromise between the various functions it performs. While gills are very efficient in taking up oxygen from water, except for a few especially adapted air-breathing fish, the secondary lamellae soon dry up and become impermeable to gases once exposed to air. Furthermore, they cohere (due to surface tension) and collapse (under gravity), more or less in a manner similar to that of shore weed at low tide. This drastically reduces the respiratory surface area. Although exposed to air (a medium rich in oxygen), paradoxically the fish becomes anoxic and hypercapnic, eventually succumbing to asphyxia. In addition to generating a large surface area in a confined volume (the opercular space), the stratified construction of the branchial system provides mechanical support that averts collapse. Typically, teleost fish have four gill arches that give rise to hundreds of gill filaments that in turn generate thousands of secondary lamellae [Figs. 11, 12]. During growth, the number of gill filaments increases more rapidly than that of the secondary lamellae. Bone and cartilage provide support to the gill arches and filaments [Fig. 12]. Such structures are lacking in the secondary lamellae where support is provided by water. In the single circulation pattern of fish, a single ventricle receives venous blood from the body and pumps it to the gills for oxygenation.

Fish can regulate the respiratory and osmoregulatory functions of gills. Such adjustments optimize the two processes to suit prevailing needs and circumstances. The mechanism has been termed "osmorespiratory compromise". It entails conflicting requirements. To effect it,

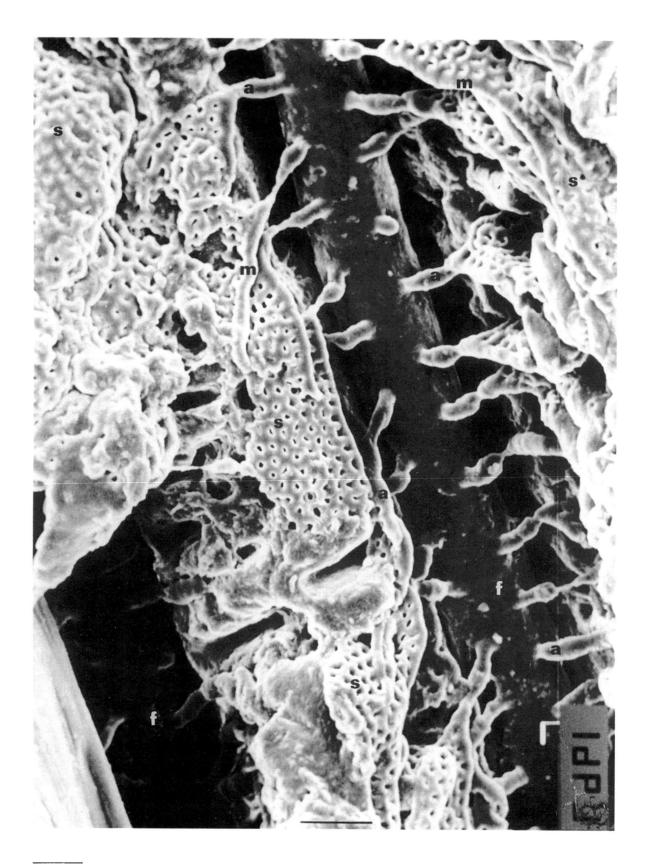

Fig. 20 Scanning electron micrograph of a preparation of injection of rubber (latex) into the blood vessels (cast) of gills of the teleost *Alcolapia grahami* showing filamental arteries (f) giving rise to secondary lamellae (s). a, afferent arterioles; m, marginal channel. Scale bar, 0.3 mm.

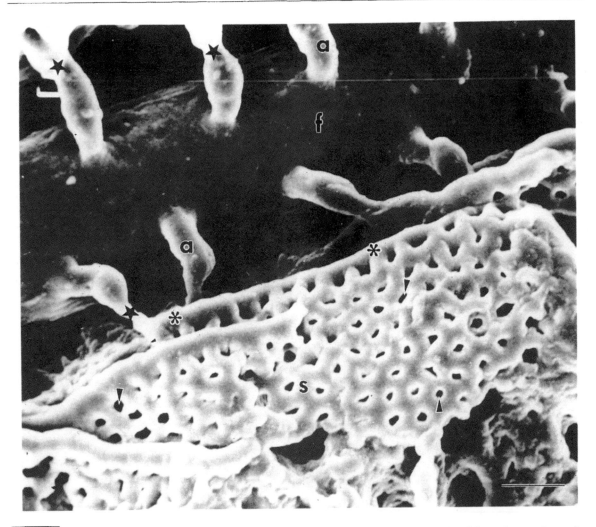

Fig. 21 Scanning electron micrograph of a perparation of injection of rubber (latex) into the blood vessels (cast) of a secondary lamella (s) of gills of the teleost *Alcolapia grahami*. a, afferent arterioles; stars, possible sphincteric sections of the afferent arterioles; f, filamental artery; asterisk, marginal channel; arrowheads, locations of pillar cells. Scale bar, 0.02 mm.

certain compromises and trade-offs are involved. Enhancing respiration by increasing the respiratory surface area results in osmoregulatory complications due to excess ionic flux, especially in fish that live in osmotically dynamic environments. Increased oxygen consumption, for example during exercise, is accompanied by increased flux of sodium (Na^+) ions. Fish can increase gas exchange (without increasing respiratory surface area) by increasing ventilation and perfusion of the gills, processes that increase the PO_2 between water and blood. The highly energetic species of fish increase the respiratory surface area of the gills while the less active ones utilize hemodynamic and ventilatory adjustments to effect osmoregulatory compromise. At rest, in the rainbow trout *Onchorhynchus mykiss*, only 60% of the secondary lamellae are perfused. The respiratory surface area of the fish gills can be maximized by altering the position of the gill filaments through contraction of the smooth muscles at their base. Adrenergic nerves enervate the muscles.

The various functions of gills occur at specific sites of a well-differentiated epithelium. Respiration and elimination of ammonia occur in the less elaborate secondary epithelium while osmoregulation and metabolism as well as regulation of pharmacologically active factors (e.g. modification of the plasma hormones in the arterial blood before they pass to the systemic circuit) occur in the primary epithelium. To avert physical damage, the gills of fast-swimming fish, such as tuna and mackerel, have developed supportive modifications. These

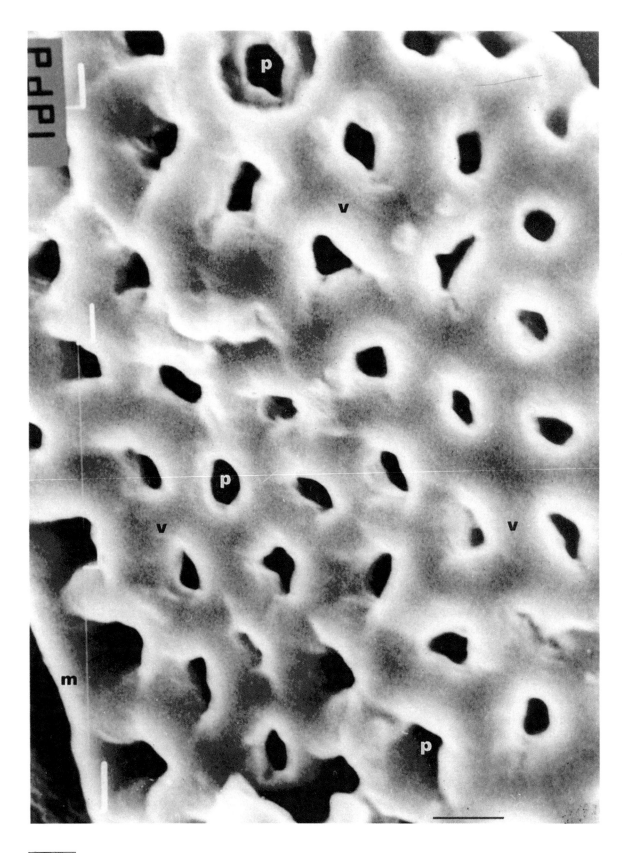

Fig. 22 Scanning electron micrograph of a preparation of injection of rubber (latex) into the blood vessels (cast) of a secondary lamella of gills of the teleost *Alcolapia grahami*. v, vascular channels; m, marginal channel. p, sites occupied by pillar cells. Scale bar, 0.01 mm.

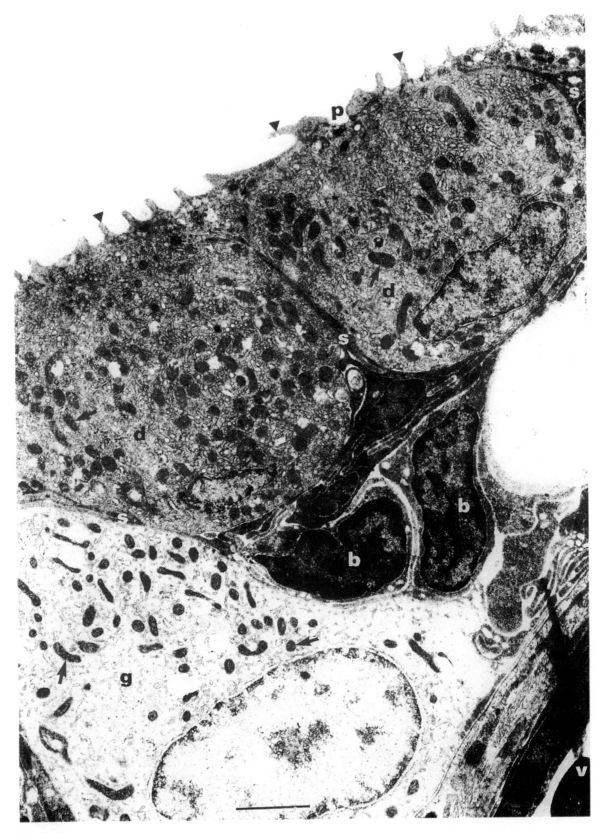

Fig. 23 Transmission electron micrograph of dark (d) (superficial) and light (g) (deep) chloride cells of the gills of teleost *Alcolapia grahami*. Cells with numerous mitochondria (arrows) and dense microtubular network. They form cellular continua from the surrounding water to the vascular channels (v). arrowheads, pavement cells with microfolds; b, basal cells; s, supporting cells; p, pore exposing a chloride cell to water. Scale bar, 15 μm.

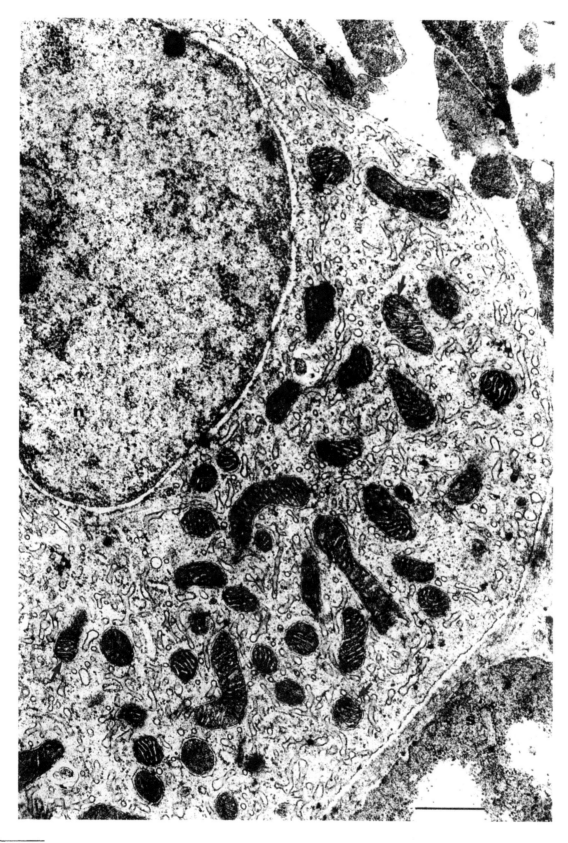

Fig. 24 Transmission electron micrograph of a chloride cell of the gills of the teleost *Alcolapia grahami*, showing an eccentrically placed heterochromatic nucleus (n) and cytoplasm containing numerous mitochondria (arrows) interspersed between a profuse network of membrane bound tubules. s, supporting cells. Scale bar, 5 μm.

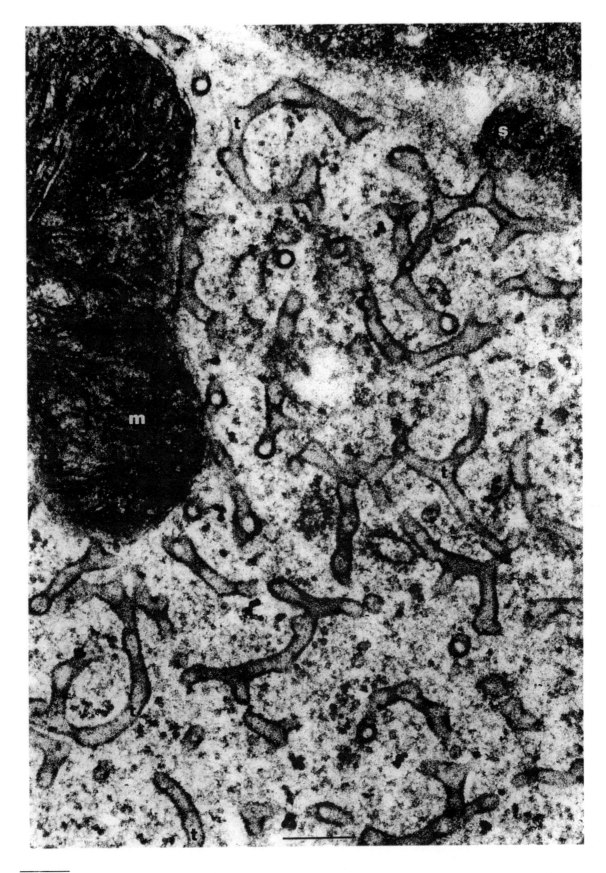

Fig. 25 Transmission electron micrograph of cytoplasm of a chloride cell of the gills of a freshwater teleost, *Oreochromis niloticus,* showing a mitochondrion (m) and microtubular network (t). s, supporting cell. Scale bar 0.1 μm.

include calcified flattened gill rays and fusion of the gill filaments. The bowfin *Amia calva*, a freshwater fish that differs greatly in behavior and habitat from marine fish, has independently acquired fused lamellae. Ram force ventilators, such as tuna and mackerel, have lost the use of the bucco-opercular pump: they have to be constantly on the move to adequately ventilate their gills.

3

Skin

1 THE SKIN: IS IT AN ARCHAIC OR A NOVEL RESPIRATORY STRUCTURE?

Relinquishment of the skin as a respiratory organ was a vital trade-off by amphibious vertebrates in their quest for colonization of the terra firma. While water loss is not a concern of aquatic animals, adequate keratinization of the skin and development of impermeable (waterproof) surface modifications such as scales was necessary to avert threat of desiccation on land. In reptiles, birds, and mammals, the skin was rendered virtually impervious to oxygen. In the human being, although after the lungs the skin presents the most extensive surface area exposed to air, only 0.2% of the total amount of oxygen is acquired through it. The skin itself utilizes the small quantity that it takes up! Only after successfully switching carbon dioxide clearance to the lung could a transitional breather exclude the skin from a meaningful respiratory role. In preparation for terrestrial life, the reptiles perfected the amniote egg, established an efficient lung, and subsequently laid down a waterproof epidermis. These features were passed on to their offshoots, the bird and the mammal. Arguably, cutaneous respiration is thought to be a derived condition unique to modern amphibians with no relationship to the Devonian ancestral lineage of the contemporary tetrapods. Moreover, it has been suggested that the Carboniferous amphibians were heavily armored. If that is correct, the highly gas and water permeable skin of modern amphibians should be an adaptative, secondarily derived condition. By holding on to the skin as a respiratory organ and water for reproduction, amphibians were consigned to wet or humid, nondesiccating habitats. For this particular reason, they are largely an inconspicuous, vertebrate group.

Cutaneous respiration presents very few possibilities for innovative change. For this reason, the skin has been dismissed as an evolutionary "dead end" or a "failed experiment". During transition from water to land, the skin played an important "bridging role" in respiratory acid-base regulation of pH as air-breathing organs progressively developed in complexity and efficiency. With gradual deemphasis of gills as a respiratory organ, ionic balance and carbon dioxide elimination were transferred to the skin. Subsequently, on attainment of terrestriality, osmotic regulation was transferred to the kidneys and clearance of carbon dioxide to the lungs. Strictly speaking, the skin is not an inert gas exchanger. The blood flow to the mid-dorsal skin of certain frogs is 1.8 times that to the ventral thoracic skin. Through movement or placement of the body on the path of a water current, passive ventilation is effected. That enhances the diffusing capacity of the skin for oxygen. Gas exchange can be altered to suit environmental conditions and circumstances. Regional dilation and constriction of the cutaneous vasculature is used to regulate perfusion. In *Bufo marinus* and *Rana catesbeiana*, dehydration leads to increased vascular resistance due to a hemoconcentrating effect. Dehydration in the terrestrial frog *Eleutherodactylus coqui* and the aquatic *Xenopus laevis* lowers the capacity to utilize aerobic metabolism during activity. The partial pressure of carbon dioxide (PCO_2) in the blood of the lungfish, *Protopterus aethiopicus* declines after vasodilation of cutaneous vasculature due to accelerated removal of carbon dioxide. By adjusting the perfusion and surface area of the skin, the hairy frog *Astylosternus robustus* can control gas exchange. In the contemporary bimodal breathers, the gill-skin system removes about 76% and the lung 24% carbon dioxide. Oxygen uptake varies with species, the respiratory structure utilized, and the habitat occupied. In the aquatic amphibians *Siren lacertina* and *Amphiuma means*, the lung takes up 65% of the oxygen needs but the gills and skin void nearly 75% carbon dioxide.

2 VERSATILITY OF THE SKIN AS A GAS EXCHANGER

As a practical rule, in natural and human engineering, design for high level of operation impels structural complexity that in turn occasions functional inflexibility. On the other hand, simplicity engenders functional multiplicity. These principles apply well in biological design. The skin, a highly malleable respiratory structure, falls in the latter design category. To varying extents, the skin is used as a respiratory organ by many animals. Compared with the ventilated gas exchangers, cutaneous respiration is a much less energy-expensive mode of acquiring oxygen. This compensates for its inherent functional and structural limitations. The water/air-blood barrier of the skin is exceptionally thick. Its diffusing capacity for oxygen is very poor.

Although fish predominantly procure oxygen through the gills, the majority of such fish acquire a substantial fraction of oxygen across the skin and similarly eliminate a significant amount of carbon dioxide through it. Many air-breathing fish have few poorly developed scales and well-vascularized skin across which substantial gas exchange can occur. Contrarily, in the eel, trout, and tench, substantial transcutaneous uptake of oxygen occurs. At a temperature of 7°C, in the common eel *Anguilla vulgaris*, buccal and cutaneous gas exchange entirely support metabolic needs. The buccal cavity of the eel is highly diverticulated and profusely vascularized. In fish larvae, before gill development, the surface of the body serves as the sole respiratory pathway.

Aquatic nonpiscine vertebrates utilize the skin as a respiratory organ. During normoxic diving, the soft-shelled turtle *Trionyx spiniferus asperus* and musk turtle *Sternotherus odoratus* can remain submerged under water for about 100 days while maintaining normal acid-base status. Many aquatic salamanders utilize a trimodal respiratory strategy: pulmonary, branchial, and cutaneous pathways variably participate in gas exchange. In the *Siren*, at 25°C, of the total amount of oxygen acquired, the gills acquire only 2.5% while eliminating 12% of the total carbon dioxide. In the cold polar water where oxygen concentrations are high, to enhance gas exchange across the skin, a fish needs only to move to adequately "ventilate" the skin. Compared with sympatric fish that have hemoglobin in the blood, the gills of the hemoglobinless Antarctic icefishes *Chaenocephalus aceratus*, *Chamsocecephalus esox*, and *Chaenichthys rugosus* have fewer secondary lamellae and the skin is relatively more highly vascularized. The resting oxygen consumption of the icefish is one-half to one-third that of hemoglobin-possessing fish of similar size. Other adaptations in the icefish include high blood volume, amounting to 7.5 % of the body weight compared to 2 to 3% in common fish. They also have a lower blood viscosity due to lack of erythrocytes.

Elimination of carbon dioxide through nonpulmonary pathways is substantial in many aquatic reptiles but only a small fraction of the total oxygen need is acquired that way. As much as 65% of the total carbon dioxide output in the aquatic turtle *Trionyx mucita* and 94%

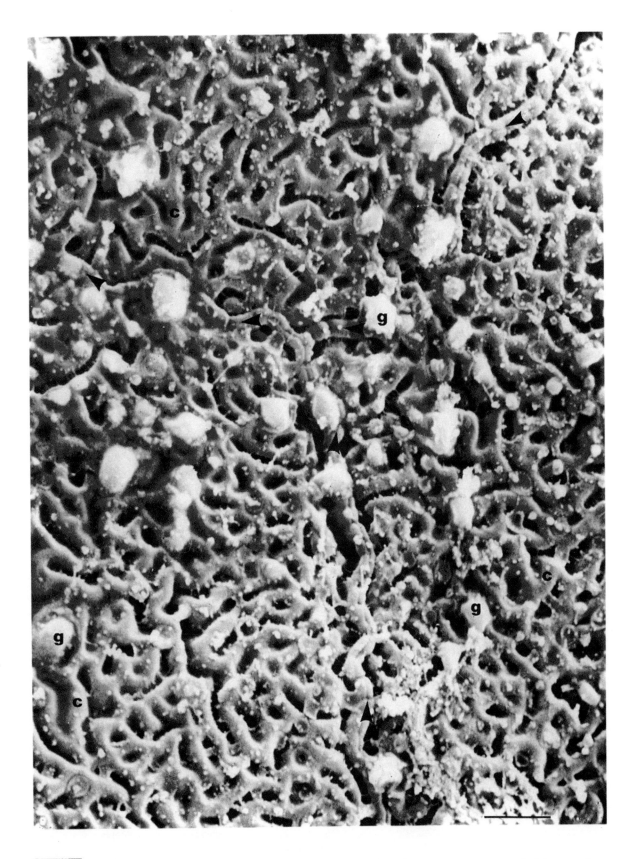

Fig. 26 Scanning electron micrograph showing vascularization of the skin of the tree frog *Chiromantis petersi*. c, blood capillaries; g, goblet cells; arrowheads, intercellular junctions. Scale bar, 30 μm.

Fig. 27 Scanning electron micrograph showing blood capillary network (c) in the skin of the tree frog *Chiromantis petersi*. Scale bar, 15 μm.

Fig. 28 Scanning electron micrograph showing folds (f) of the mucosal lining of the buccal cavity of the teleost *Alcolapia grahami*. Scale bar, 0.14 mm.

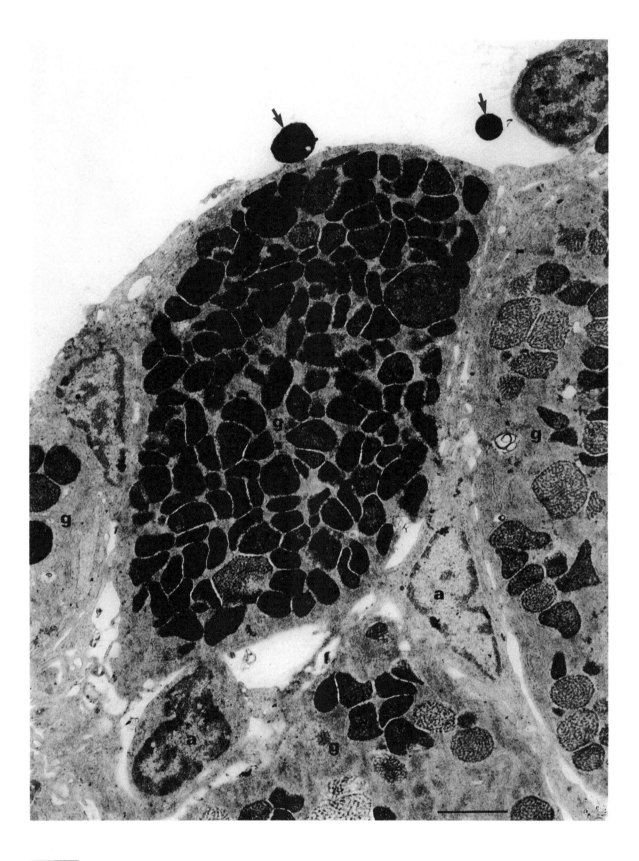

Fig. 29 Transmission electron micrograph showing mucus secretory cells (g) of the epithelial lining of the buccal cavity of the teleost *Alcolapia grahami*. a, interstitial cells or immature epithelial cells; arrows, secretory granules. Scale bar, 0.75 µm.

in the sea snake *Pelamis platurus* is lost cutaneously. In the aquatic snake *Acrochordus javanicus*, 33% of carbon dioxide elimination and only 8% of the total oxygen uptake occurs through the skin. Inhabiting cold well-oxygenated waters, the urodele plethodontid salamanders that constitute the largest family among the Caudata, lack lungs and gills: they rely entirely on a highly vascularized skin for respiration. The length of the skin blood capillaries constitutes 90% of all blood vessels associated with the respiratory surfaces, with the remaining 10% being in the buccal cavity. In this respect, it is worth noting that plethodontid salamanders are thought to have originated from torrential mountain streams in which the water is very well aerated. Adaptatively, they have adopted long cylindrical body shapes that increase the external surface area. Movements that stir the external respiratory medium and subcutaneous perfusion act as the only aids to the diffusive conduction of the skin for oxygen. In amphibians, those species with smooth, poorly vascularized lungs have a thinner epidermis and a much denser dermal capillary network than those with well-developed lungs. The relative length of cutaneous capillaries expressed as percentage of the total length of respiratory surface-associated capillaries is about 50 to 80%. On exposure to extremely hypoxic or hypercarbic water, the totally aquatic and sorely cutaneous-breathing amphibians *Cryptobranchus alleganiensis* and *Telmatobius culeus*, rock or sway their bodies back and forth. High-altitude hypoxia does not appear to restrict the spatial distribution of amphibians: eleven species of frogs live between elevation of 3.7 and 5.2 km. *Telmatobius culeus*, a frog found in Lake Titicaca (Bolivia-altitude 3.8 km) is adapted to low aquatic PO_2. Its skin is well vascularized and folded while the blood has small, numerous erythrocytes (the highest count among amphibians), high hemoglobin concentration, and high hematocrit. When swimming, the frog increases the convective movement of water over the skin through an unusual "bobbing" behavior characterized by violent agitation of the water. Interestingly, the skin of the tree frog *Chiromantis petersi*, a highly xerophilic frog (it seeks out dry usually hot environments) is highly vascularized [Figs. 26, 27]. The surface of the buccal cavity of the teleost fish. *Alcolapia grahami* is folded, poorly vascularized [Fig. 28], and is lined by a secretory epithelium [Fig. 29].

4

Swim (Air) Bladder

1 EVOLUTION

All bony fish (Osteichthyes) have a swim (air) bladder, at least at one stage of their development. From evolutionary, structural, and functional perspectives, among vertebrates, few organs have attracted and sustained as much interdisciplinary scientific inquiry, debate, and controversy as the swim (air) bladder of fish. The biology of the swim bladder is of particular interest for the following reasons: a) in some fish it serves as an accessory respiratory organ and hence its structure and function may shed some light on the evolution of respiratory organs and stratagies; b) its role as a hydrostatic organ may help explain the adaptive processes that were essential for subsistence in water (a fluid medium of high viscosity)—of particular interest are the postural and locomotory adjustments that ensued for habitation on land where gravity impacted more significantly on the size and shape of animals; and c) the capacity of the organ to secrete and concentrate gases (especially oxygen) to several hundreds of atmospheres of pressure while living in water (a medium inherently deficient in oxygen) is a unique biophysical phenomenon.

The question—did the swim bladder produce the lung or vice versa and are the two organs homologous or analogous?—has long been passionately debated. The issue is far from settled. It is generally thought that the lung evolved in freshwater vertebrates during the Paleozoic Era as an adaptation to prevailing hypoxia. While this line of reasoning is intuitively appealing, to the contrary, subsistence of pelagic marine fish in well-oxygenated waters signifies a predominant hydrostatic function of the swim bladder. Moreover, it has been argued that swim bladders developed to offset increase in weight with formation of a bony skeleton in fish. The lungs of lungfishes (Dipnoi) and the ancestral actinopterygian fish such as *Amia calva*, *Polypterus bichir*, and *Lepisosteus osseus,* are generally considered homologous to swim bladders.

2 DEVELOPMENT

In teleosts, Dipnoi, and Polypteridae, the swim bladder and the lungs develop as outgrowths from the wall of the foregut, with the primodial air-sac/lung becoming evident very early. Branches of the vagus (cranial nerve) and the sympathetic nerves enervate both organs. That, however, is as far as the similarities go. The differences between the two organs are:

a) the swim bladder arises from the dorsal or lateral walls of the foregut while the lungs arise from the ventral aspect (i.e., floor); b) the lungs receive left and right pulmonary arteries that originate symmetrically from the last pair of epibranchial arteries while the swim bladder is supplied with arterial blood directly from the aorta; c) the lungs remain permanently connected to the pharynx while only physostomous (i.e., connected to the alimentary canal) swim bladders are connected to it (the physoclistous are not); and d) the lungs are lined with a surface active factor (surfactant)—a complex mixture of phospholipids, neutral lipids, and proteins—which apparently is lacking in swim bladders. Given their ventral location and the inherent problems regarding balance that would accompany such a topographically placed organ, it is conceivable that the lungs may have invariably been solely respiratory while the dorsally located swim bladder could combine hydrostatic and respiratory roles. Moreover, use of a respiratory organ for hydrostatic control is inefficient because when oxygen is removed and little if any carbon dioxide is secreted back, the animal becomes less buoyant and sinks. In such a case it is imperative that removal of oxygen from the lung be synchronized with hydrostatic adjustment. A fish with a physostomous swim bladder used in gas exchange has to make regular sojourns to the surface to fill the bladder/lung. Energetically, this is a costly process. In teleosts, the swim bladder is mainly hydrostatic. The organ is structurally much less complex than the Dipnoan and Polypteridae lungs that are mainly respiratory. In amphibians, the nonseptate saccular lungs, for example of urodeles and salamanders such as *Cryptobranchus* and *Necturus* are predominantly hydrostatic.

Based on the present morphogenetic data, the hydrodynamic and respiratory roles of the swim bladder are by no means mutually exclusive and neither can the possibility that the two organs evolved independently be totally disregarded. In fact, some modern fish demonstrate that a nascent air-containing organ may have played both roles. Teleosts in which the swim bladder serves as a lung include *Arapaima* (Osteoglossidae), *Gymnarchus* (Mormyridae), *Erythrinus* (Characinidae), *Umbra* (Esocidae), (Notopteridae), and *Lepidosteus* (*lepiostidae*). The tarpon, *Notopterus Megalops* regularly ventilates its physostomous swim bladder that opens to the mouth even in normoxic water. From its alveolar-like tissue, the tarpon's swim bladder is like a lung. In *Erythrinus unitaeniatus*, the middle region of the physostomous bladder is well vascularized and significantly utilized for gas exchange. The swim bladder of the teleost *Alcolapia grahami*, a small cichlid fish that lives in the highly alkaline lagoons of the volcanic Lake Magadi of Kenya, has a well-vascularized swim bladder [Figs. 30, 31]. The bladder is particularly used for respiration after strenuous exercise and especially during the night when the water is virtually anoxic.

3   LOCATION AND STRUCTURE

There are two types of swim bladders: the physostomous ones are connected to the gastrointestinal system (pharynx) (Fig. 32) while the physoclistous ones are not, i.e. they are closed off. In general, surface dwelling fish have physostomous swim bladders while the deep water dwelling ones have physoclistous ones.

From the outer to the inner surface, the wall of the swim bladder comprises: a) a thin squamous epithelial cell layer; b) a large space containing collagen, amorphous ground substance, and elastic tissue; c) a band of smooth muscle; d) a space containing connective tissue; and e) gas-gland cells that protrude into the airspace [Figs. 33, 34]. The tight packing of the tissue layers of the wall of the swim bladder and the presence of guanine crystals are important structural features that prevent loss of gases by back-diffusion. In some swim bladders, the gas-gland cells form a single layer of columnar cells [Fig. 33, 35]. On the juxtaluminal aspect, the cells join across conspicuous tight-junctions (zonula occludens) and much deeper across desmosomes (macula adherens) [Figs. 36-38]. Complex (multilayered) gas-gland epithelia are found in the euphysoclistous swim bladders, i.e., those ones that do not open to the mouth, of fish such as the gadonids, *Motella* and *Gadus,* in the horse mackerel, *Trachurus*, the trigger-fish, *Balistes*, and deepsea fish in general.

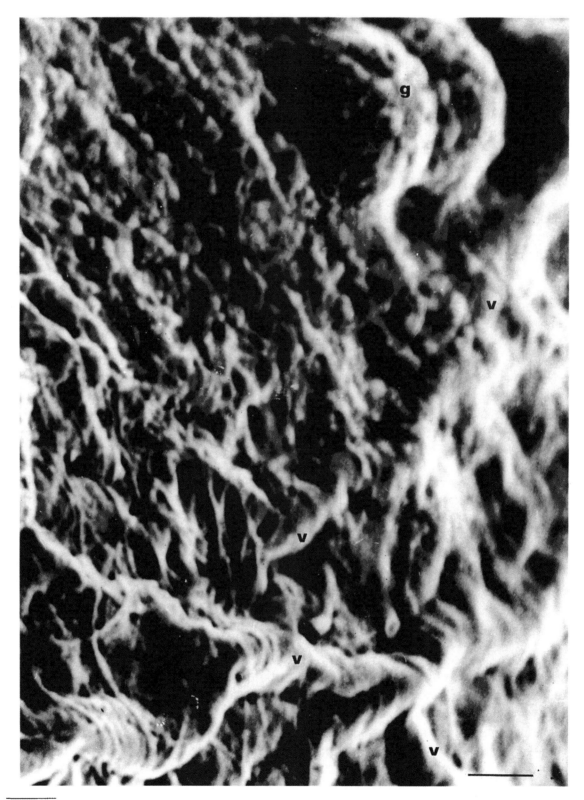

Fig. 30 Scanning electron micrograph of the luminal surface of the swim bladder of the teleost *Alcolapia grahami* showing blood vessels (v) and gas-gland cells (g). Scale bar, 8 μm.

The secretory efficiency and composition of the gases produced by the swim bladder correlate with the structural complexities of the gas-gland epithelium and the rete mirabile a structure that affords a counter-current exposure of the afferent blood, i.e., blood to the swim

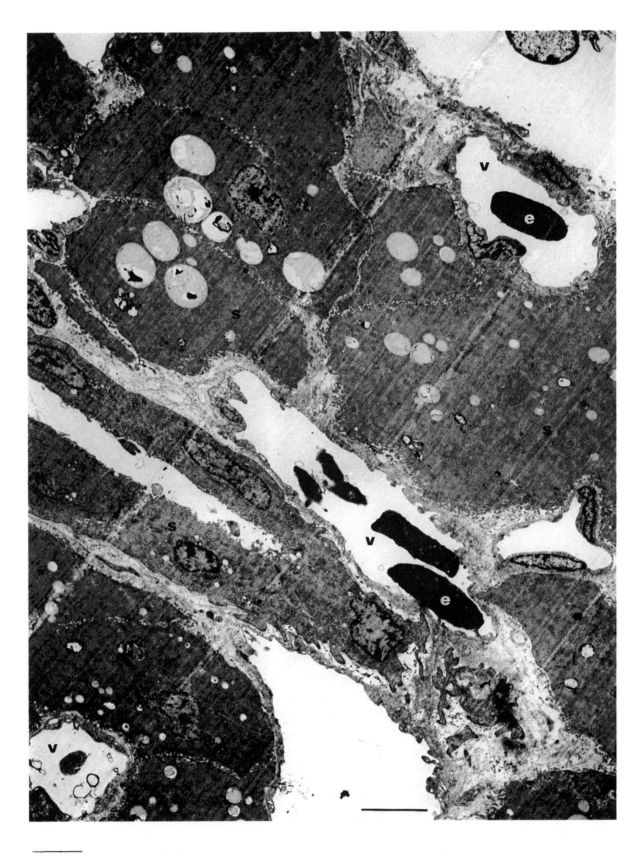

Fig. 31 Transmission electron micrograph of a cross-section of the wall of the swim bladder of the teleost *Alcolapia grahami* showing blood vessels (v) and smooth muscle tissue (s). e, erythrocytes. Scale bar, 3.5 μm.

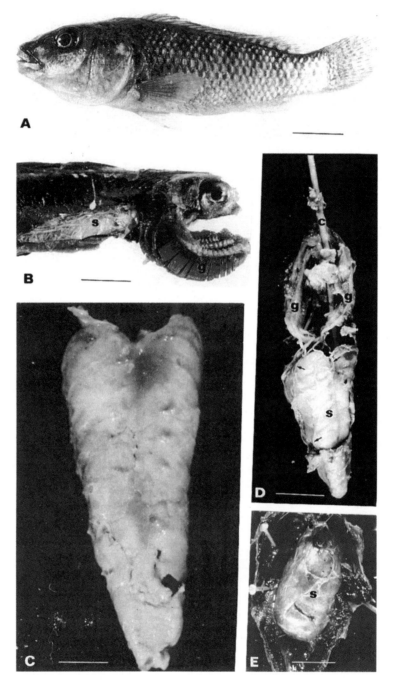

Fig. 32 A: Photograph of *Alcolapia grahami*, a small tilapiine cichlid fish which lives in the highly alkaline Lake Magadi of Kenya where there are dramatic diurnal fluctuations of levels of oxygen in water. B: Photograph showing the location of the swim bladder (s) of *Alcolapia grahami*. The organ is located posterior to the gills (g) on the dorsal aspect of the body truck. C: Photograph of dorsal view of cast of the swim bladder of *Alcolapia grahami* prepared by injection of latex rubber into the organ. D: Photograph of a cast of the blood vessels of gills (g) and swim bladder (s) of *Alcolapia grahami* prepared by injection of latex rubber. The cannula (c), was placed to show that the swim bladder opens into the mouth, i.e. the swim bladder is physostomous. Note that the blood vessels of the gills (arrows) are connected to those of the swim bladder indicating that the organ can serve as a respiratory organ. E: Photograph of a swim bladder (s) of *Oreochromis niloticus*, a fresh water teleost. The bladder is more rounded in shape and is located on the roof of the coelomic cavity. A: Scale bar, 1 cm; B: Scale bar, 1 cm; C: Scale bar, 0.2 cm; D: Scale bar, 0.5 cm; E: Scale bar, 0.5 cm.

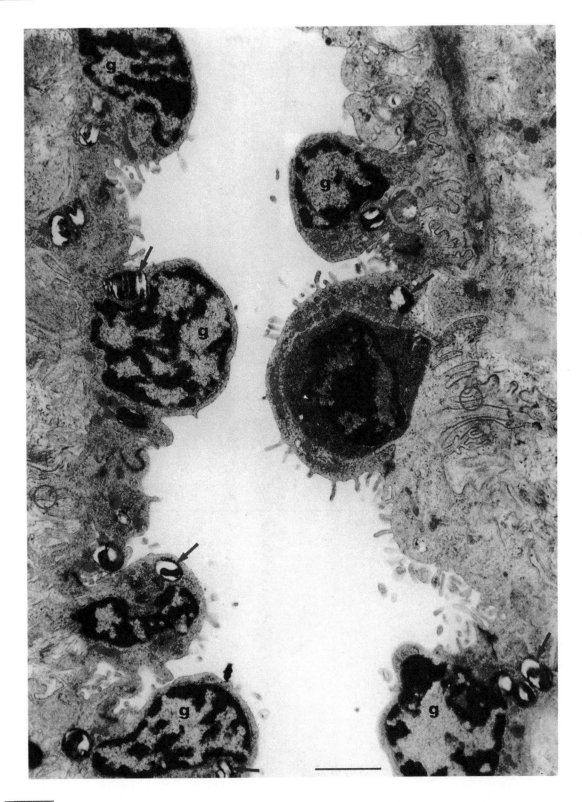

Fig. 33 Transmission electron micrograph of gas-gland cells (g) lining the wall of the swim bladder of the teleost *Alcolapia grahami*. Note amplifications of the basal membranes (circles). Arrows, secretory bodies; s, smooth muscle. Scale bar, 2.7 μm.

bladder and efferent blood, i.e., blood from the swim bladder. Compared to surface-dwelling teleosts, bathypelagic teleosts have relatively more elaborate gas-gland epithelia and rete mirabile. Topographically, the gas-gland cells are located close to the retial blood capillaries

Fig. 34 Transmission electron micrograph of the longitudinal section across the wall of the swim bladder of the teleost *Alcolapia grahami* showing relationship between smooth muscle (s) and basal cell membrane amplifications (circles) of the gas-gland cells. c, collagen fibers. Scale bar, 1.3 μm.

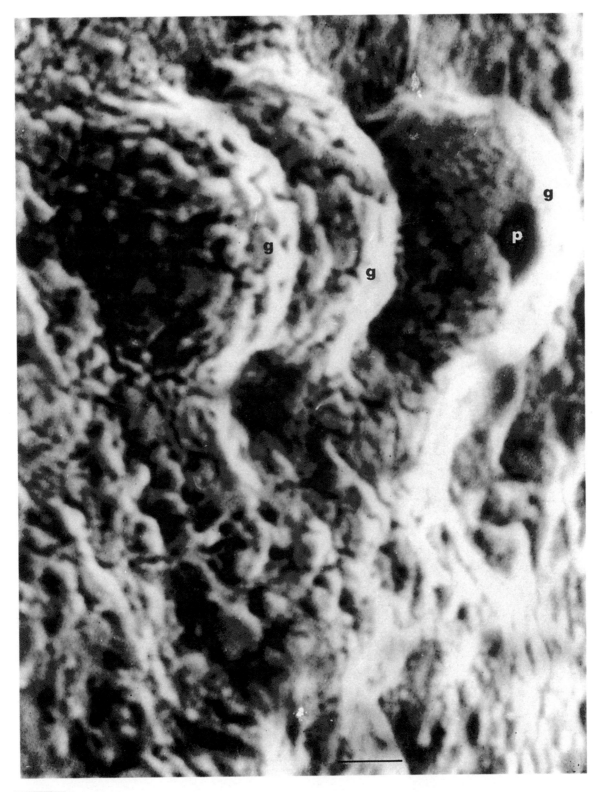

Fig. 35 Scanning electron micrograph showing the rough-surfaced gas-gland cells (g) of the swim bladder of the teleost *Alcolapia grahami*. p, secretory pit. Scale bar, 15 μm.

[Fig. 30]. Some of the gas-gland cells have remarkably long microvilli [Fig. 33, 37], others a ruffled surface, while some have a fairly smooth outer membrane [Figs. 36, 38]. It is conceivable that various populations of gas-gland cells line the luminal surface of the swim bladder or that the cells may change greatly with the stage of development, age, and secretory states. Gas-gland cells have large, highly euchromatic nuclei, and dense

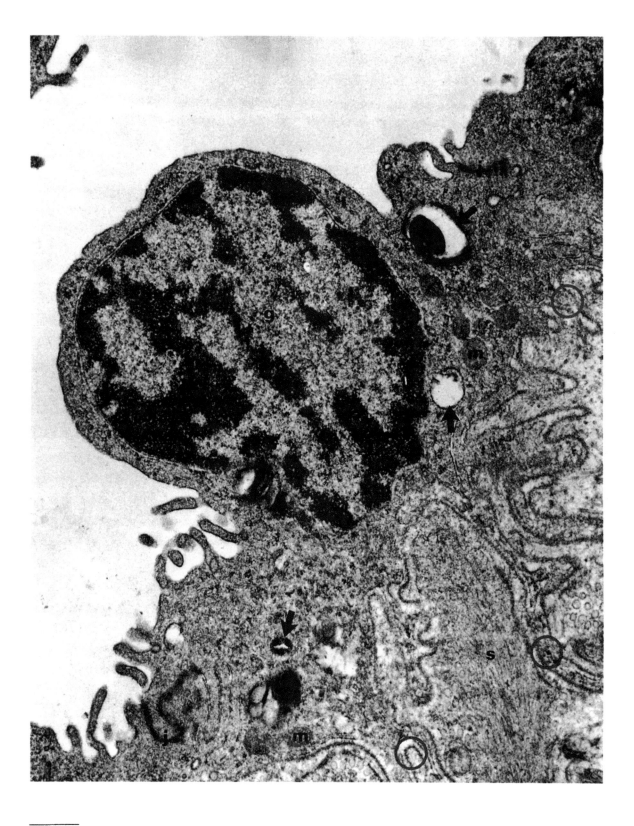

Fig. 36 Transmission electron micrograph showing smooth-surfaced gas-gland cell (g) of the swim bladder of the teleost *Alcolapia grahami* with secretory bodies (arrows) and amplifications of the basal cell membrane (circles). s, smooth muscle; j, intercellular cell junction; m, mitochondria. Scale bar, 1.6 μm.

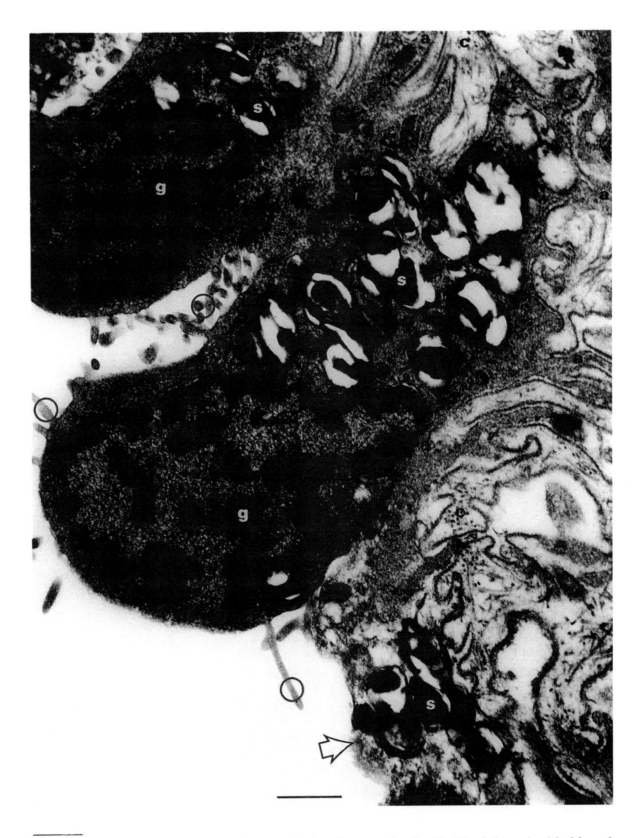

Fig. 37 Transmission electron micrograph showing gas-gland cells (g) of the swim bladder of the teleost *Alcolapia grahami* with microvilli (circles), abundant secretory bodies (s), and basal cell membrane amplifications (a). Arrow, a putative apoptotic gas-gland cell; c, collagen fibers. Scale bar, 1 µm.

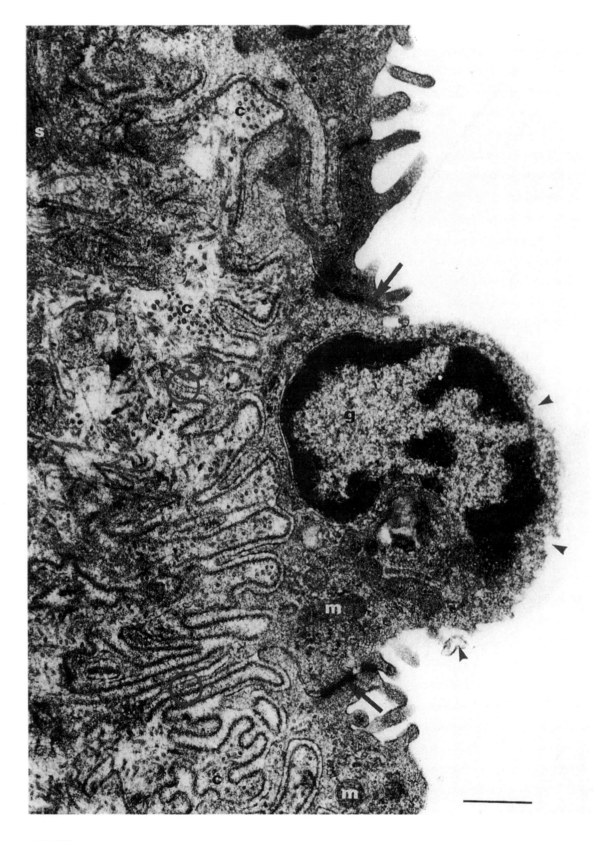

Fig. 38 Transmission electron micrograph showing gas-gland cell (g) of the swim bladder of the teleost *Alcolapia grahami* with a ruffled surface over the perikaryon (arrowheads). Arrows, intercellular junctions. The gas-gland cell is attached to a subepithelial space that consists of smooth muscle (s) and collagen fibers (c). The basal cell membrane is highly amplified (circles). m, mitochondria. Scale bar, 0.75 μm.

intracytoplasmic bodies [Figs. 33, 36-38]. In the cod, *Gadus,* lipid-like material (possibly phospholipid in nature) covers the luminal surface of the swim bladder. It is not clear, however, whether the substance is surface active or not. Mitochondria are scattered in intensely granulated cytoplasm. On the basal aspect (i.e., the side fronting the retial blood capillaries), the cell membrane of the gas-gland cells is highly folded [Figs. 33, 36, 37-39]. The extensions profusely interdigitate with the underlying connective tissue elements. The basal cell membrane amplifications may enhance exchange of materials between the gas-gland cells and the underlying structural components, especially the retial blood capillaries. Moreover, the firm attachment of the gas-gland cells to the underlying connective tissue elements of the wall of the swim bladder may render them the necessary mechanical support. This may be particularly important in maintaining the structural integrity of cells exposed to shifting intraluminal pressures. The retial blood capillary endothelial cells are highly sporadically attenuated [Fig. 31]. Abundant micropinocytotic vesicles occur therein. This indicates active transfer of substances between the blood capillaries, the gas-gland cells, and the surrounding tissue elements.

4 FUNCTION

With regard to energy expenditure, gas-filled cavities are the most cost-effective strategies for achieving and regulating neutral buoyancy in aquatic animals. However, alternative means for hydrostatic adjustment such as low-density tissues or organs, accumulation of lipids in the air bladder, and water in the tissues have evolved. Swim bladders are lacking in many groups of fishes such as sharks and rays (elasmobranchs). Where they occur, swim bladders perform multiple functions. The best known functions are buoyancy regulation, gas exchange, hearing, sound production, and pressure sensation.

Molecular oxygen constitutes the greatest proportion (up to 95%) of the gas mixture secreted into the swim bladder of fish, with nitrogen, carbon dioxide, and rare gases such as argon constituting relatively smaller proportions. The PO_2 in the swim bladders of fish that live at great depths may exceed 150 atmospheres. In such cases, the PO_2 in the swim bladder is as high as 10,000 times that in the surrounding water. Considering that all the gases must in the first instance be derived from water where they occur at very low concentrations, the high pressure in the swim bladders shows the remarkable secreting and concentrating efficiency of the organ. In the European eel *Anguilla anguilla,* 83% of oxygen extracted from water is discharged into the swim bladder; only 17% is used for metabolic purposes.

A multiplicative process in the countercurrent system of the rete mirabile is thought to explain the secretion of gases into the swim bladder. In physoclistous swim bladders, lactic acid produced through glucose metabolism in the gas-gland cells is secreted into the afferent retial blood. On acidifying the blood, oxygen is forced out of it by the Bhor effect (i.e, pH-dependent release of oxygen from the hemoglobin), concentrating it in the swim bladder. PCO_2 is increased by conversion of bicarbonate ions to carbon dioxide while that of nitrogen and inert gases is raised by a reduction in their physical solubility, i.e. "salting-out effect". Back-diffusion of carbon dioxide in the rete mirabile increases the pH of blood, enhancing secretion of oxygen into the swim bladder. Although the gas-gland cells [Figs. 31, 33, 35-38] differ little ultrastructurally from the typical epithelial cells lining the vertebrate respiratory airways [see, e.g. Figs. 86-88, 91: Chapter 9], the cells produce lactic acid even while exposed to high concentrations of oxygen. At a hyperbaric pressure of about 50 atmospheres, the gas-gland tissue of *Sebastodes miniatus* still secretes lactic acid. The remarkable efficiency of the gas-gland cells in converting glucose to lactic acid is only exceeded by that of tissues such as the retina.

In the physostomous (open) bladders, a neuromuscular mechanism is involved in the uptake and ejection of air. Buccal force pumping and expiration may effect inspiration by activity of the smooth muscle in the pneumatic duct of the bladder. In *Hoploerythrinus,* the

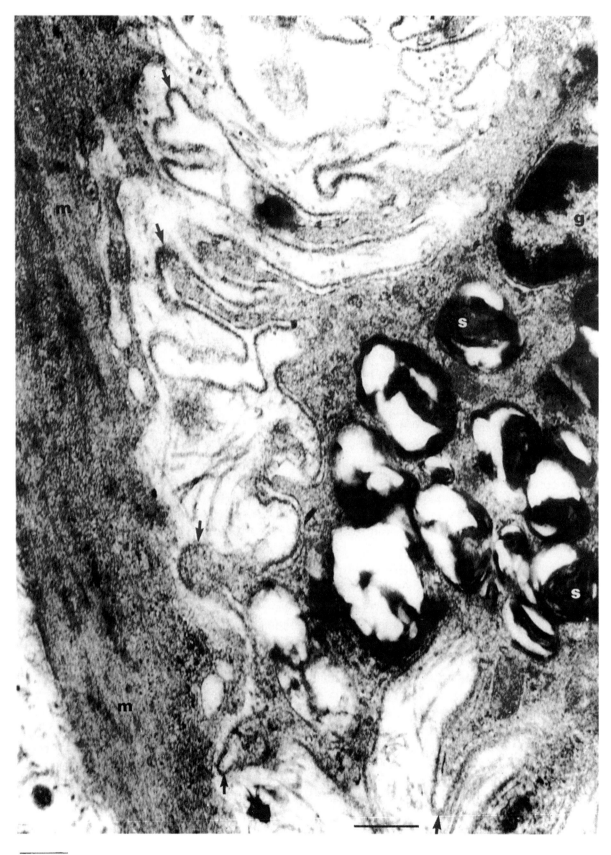

Fig. 39 Transmission electron micrograph showing basal cell membrane amplification (arrows) of a gas-gland cell (g) of the swim bladder of the teleost *Alcolapia grahami*. m, smooth smuscle; s, secretory granules. Scale bar, 0.60 μm.

pneumatic duct is particularly wide and muscular while in the eel *Anguilla anguilla*, the duct is adapted for gas exchange by having a distinct circulatory system. In most physostomous fish, e.g. in salmonids, cyprinids, electric eel, and notopterids, the entrance of the pneumatic duct to the esophagus has smooth and striated muscle sphincters. The elasticity of the bladder coupled with the contraction of the muscles of the body wall may aid in the ejection of gases from the swim bladder.

5

Transitional (Bimodal) Breathers

1 EVOLUTION OF BIMODAL BREATHING

Classification of contemporary animal life into aquatic and terrestrial fauna is as old as the science of biology itself. The morphological attributes justifying this cassification have mainly been decreed by the various biophysical attributes of the fluid media, water and air, to which the animals have specifically adapted. While the conspicuous differences between terrestrial and aquatic animal forms are readily appreciated, the motley of variably adapted animals regularly traversing the air-water interface is often overlooked. To illustrate this disregard, the expression "like a fish out of water," used to describe a person in unfamiliar circumstances, conveys the notion that fish cannot survive by respiring air. While this may be true for some fish, it does not apply absolutely to the entire piscine taxa. Some suitably adapted fish breathe air and will actually drown if forcibly kept submerged under water! For example, the adult piracucu, *Arapaima gigas,* and the lungfish *Protopterus aethiopicus* die if refused access to atmospheric air, even when kept in well-oxygenated water.

Some transitional (= intermediate = bimodal = dual) breathers permanently live in water and extract oxygen from the air above. Others physically commute between air and water, while some start their development in water and end up on land at maturity. As many as 370 species of fish are known to breathe air. More than any other vertebrate group, fish display the most extensive range in the evolution of air-breathing. Modern transitional breathers use multiple structures for respiration. Biologically, the taxon constitutes a highly provocative group, representing a modern prototype of the animals that pioneered the transformation of water to air-breathing and the transition from water to land. Developing from the holosteans during the late Triassic Period, the modern teleosts are, however, too distant to be the direct ancestors of the contemporary terrestrial air-breathing vertebrates. Air-breathing is believed to have evolved as an adaptation to hypoxic or anoxic conditions in water. In the entire ichthyoid fauna, only the Holostei and the tropical freshwater teleosts are notably air-breathers. There is no paleontological evidence to show that direct emergence from water to land ever occurred in any animal taxon. It would appear that the biophysical differences between water and air and the different design and structural prescriptions decreed on the respiratory organs are too great to allow a direct shift from utilizing water to air and vice versa.

Together with the ancestral crossopterygians that gave rise to the primitive amphibians of which the sole survivor is the coelacanth, *Latimeria chalumnae*, the lungfishes (Dipnoi) are arguably the closest extant relatives of the modern tetrapods. For this particular reason, the group provokes substantial interest among biologists. Among vertebrates, the pulmonary arteries first appeared in lungfishes as branches of the sixth pair of aortic arches. Early in their development, the blood vessels supplied the swim bladder, a structure thought to have given rise to the lungs [Chapter 4]. The lungfish (a taxon in which a pulmonary vein and a partly divided heart first appear) presents a pivotal point in the evolution of double circulation. From simple diffusion dependent unicells via open circulation and single circulation, it took nearly 300 million years for double circulation to form. In the African genera of lungfish, *Protopterus*, the arteries of the embryonic third and fourth branchial arches lack gill filaments: they form shunt vessels that correspond to the carotid and systemic arches of the Amniota.

From the diversity of the evolved accessory respiratory organs and the adopted respiratory strategies, air-breathing was achieved independently by different vertebrate taxa in response to a common factor—hypoxia. Respiratory adaptations for air-breathing are found in most or all of the late Paleozoic fishes. The oxygen content of water in the Silurian Period was as low as 1.9 kPa. Air-breathing has only evolved in lineages of two bony fishes. These are the actinopterygian (ray-finned) fish [i.e. bichirs (gar, polypterus, and bowfin) and teleosts] and the sarcopterygian (lobe-finned) fish [i.e. lungfishes (Dipnoi)]. The near total extinction of the crossopterygian fishes makes understanding the evolution of respiratory processes much more difficult to grasp. It is important to underscore that **air-breathing first evolved as a direct response to prevailing lack of oxygen and not as an adaptation for terrestrial habitation.** Per se, air-breathing just happened to be an important preadaptation (among many others) for transition from water to terra firma when that became necessary. In this regard, the following fact is noteworthy: **animals cannot possibly anticipate and prepare for prospective events, especially catastrophes!**

Bimodal-breathing is estimated to have evolved at least 67 different times in the past. Inaugural lungs are found in three genera of lungfishes (Dipnoi) [the South American *Lepidosiren*, the African *Protopterus*, and the Australian *Neoceratodus* (in all six living Southern Hemisphere fresh-water species)], in chondrosteans [*Polypterus* and *Calamoichthys*], and in holosteans [*Amia* and *Lepisosteus*]. In many ways, developmentally these fish have been extremely biologically conservative. Since acquiring the capacity for air-breathing, they have changed very little morphologically. They are important "living edifices" that illustrate the adaptive strategies and the backgrounds against which air-breathing first evolved. Over the 300 million years of their existence, lungfishes such as *Protopterus aethiopicus* and *Lepidosiren paradoxa* have adapted so well to air-breathing that, as a matter of fact, contact with air is a more powerful stimulus than tactile or pain stimuli.

2   DIVERSITY AND STRUCTURE OF ACCESSORY RESPIRATORY ORGANS

Bimodal-breating fish rely to varying extents on air and water for their oxygen needs. Factors such as degree of development of the accessory respiratory organs and availability of oxygen in water cause the fish to switch from breathing water to air. The more terrestrial species are obligate air-breathers while the more aquatic ones are facultative air-breathers. Among the actinopterygian fishes, only the polypterids have developed definite lungs. The polypterid lungs are relatively more primitive than those of the Dipnoi. With the right lung better developed than the left one, the polypterid lungs are slender and lack conspicuous internal subdivision. *Lepidosiren paradoxa* and *Protopterus aethiopicus* evolved in separate geographical sites after the splitting and drift of continents in the early Mesozoic. Though spatially separated, the fish continued to live in similar ecological habitats: *Lepidosiren* is found in the slowly flowing, poorly oxygenated waters of the Amazonian basin while *Protopterus* occurs in the derelict inland fresh-water masses and seasonal rivers of Africa. *Protopterus* has paired, symmetrical, internally well-subdivided lungs [Figs 40, 41]. The lungs are so efficient and the fish so dependent on air that the gills are vestigial. Obligate air-

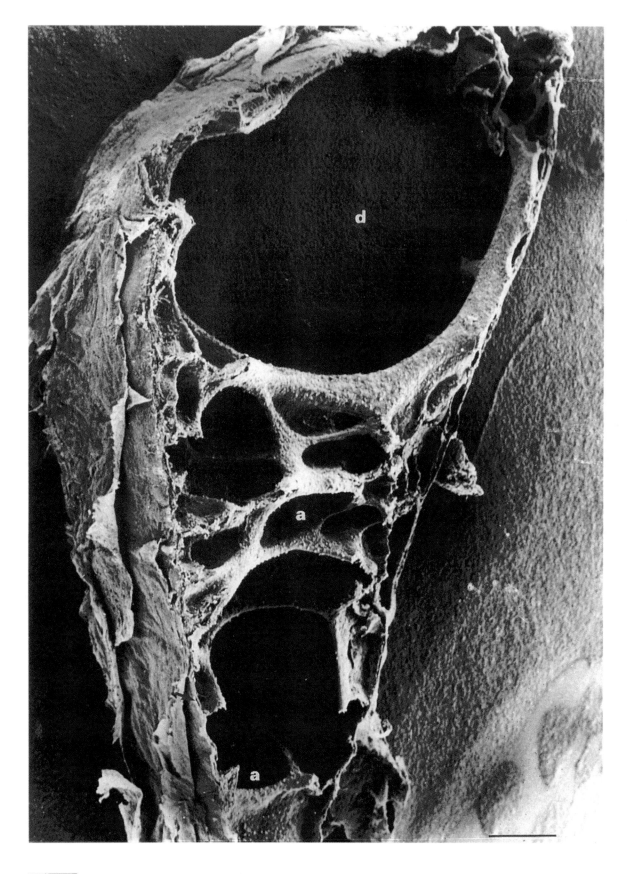

Fig. 40 Scanning electron micrograph of the lung of the lungfish *Protopterus aethiopicus* showing an eccentrically placed air duct (d) and peripherally located air cells (a). The lung is heavily invested in surface coverings (arrows). Scale bar, 0.13 mm.

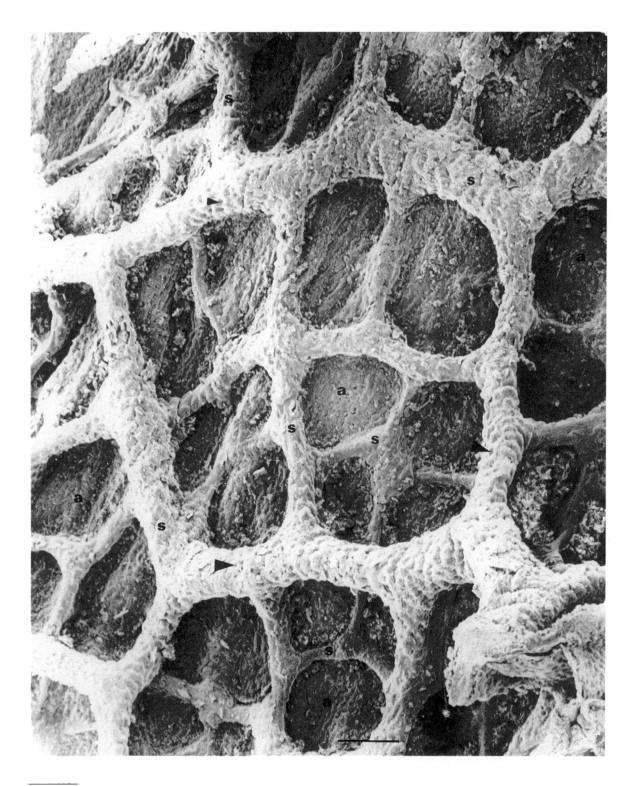

Fig. 41 Scanning electron micrograph showing the luminal surface of lung of the lungfish *Protopterus aethiopicus* showing septa (s) at different levels that subdivide the lung into air cells (a). Arrowheads, depressions in which the perikarya of the epithelial cells are located. Scale bar, 0.4 mm.

breathers, both *Lepidosiren* and *Protopterus* have a better carbon dioxide buffering capacity of blood than the Western Australian lungfish, *Neoceratodus forsteri*, itself a facultative air-breather. *P. aethiopicus* acquires 89% of its oxygen needs and eliminates 40% carbon dioxide

Fig. 42 Photograph of a gross specimen of the respiratory organs of the bimodal-breather *Clarias mossambicus* showing gill arches (g), suprabranchial chamber membranes (s), and labyrinthine organs (b). r, gill rakers; f, gill fans. Scale bar, 10 mm.

through the lungs. *Neoceratodus* has a single, unpaired, thin-walled but internally subdivided lung. It has well-developed gills. The thickness of the blood-gas barrier in the lung of *N. forsteri* ranges from 1.5 to 2.5 μm, that in the lung of *L. paradoxa* is 0.85 μm, while that of *P. aethiopicus* ranges form 0.37 to 0.85 μm; the value for the lung of *Polypterus bichir* is 1.22 μm. As the dry season approaches, *Protopterus amphibius* encases itself in a cocoon of hard

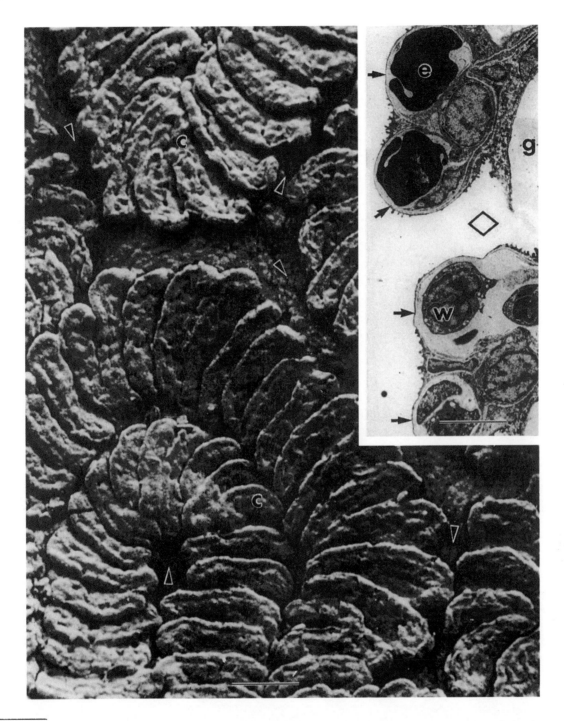

Fig. 43 Scanning (main figure) and transmission (insert) electron micrographs of the luminal surface of the suprabranchial chamber membrane of the bimodal-breathing catfish, *Clarias mossambicus* showing transverse capillaries (c) and bare tracts (arrowheads) that contain secretory goblet cells. Insert shows a cross-section of the suprabranchial chamber membrane. e, erythrocytes contained in transverse capillaries (arrows); diamond, smooth tract with underlying goblet cells (g); w, white blood cell. Scale bar, 11 μm; insert scale bar, 13 μm.

soil. Periodically breathing through a "snorkel" that opens to the surface, it has been known to survive in a semianimated state for as long as 5 years!

The accessory respiratory organs of fish present remarkable morphological diversity. The skin [Chapter, 3], the buccopharyngeal membrane, the suprabranchial chamber membranes, the labyrinthine organs, the gastrointestinal system, and the swim (air) bladders [Chapter, 4]

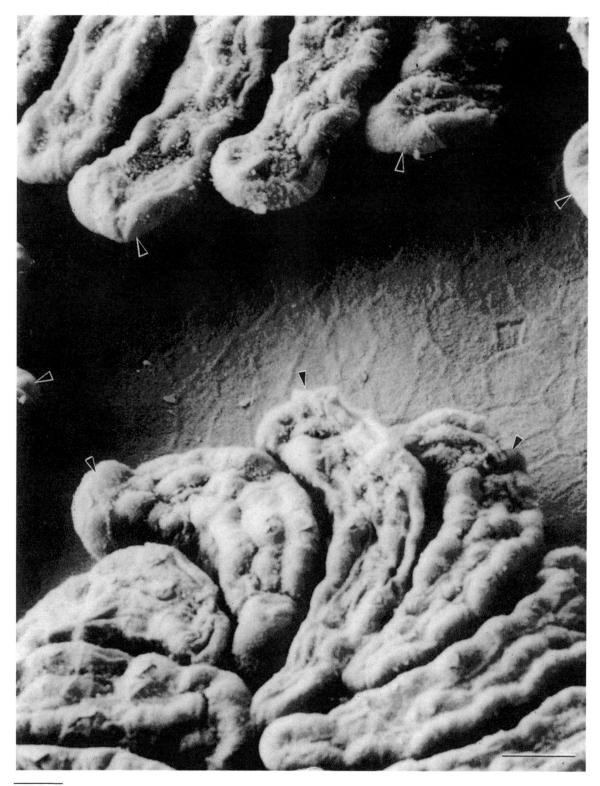

Fig. 44 Scanning electron micrograph showing luminal surface of the suprabranchial chamber membrane of the bimodal-breather *Clarias mossambicus* showing transverse capillaries (arrowheads) and bare tracts that contain secretory goblet cells. Scale bar, 11 μm.

are variably used as auxilliary respiratory organs. In *Plecostomus* and *Ancistrus* (both tropical Siluroidae), the stomach serves as a respiratory organ: fresh air is swallowed and stale regurgitated. In the pond loach, *Cobitis* (= *Misgurnus*), the middle and distal parts of the

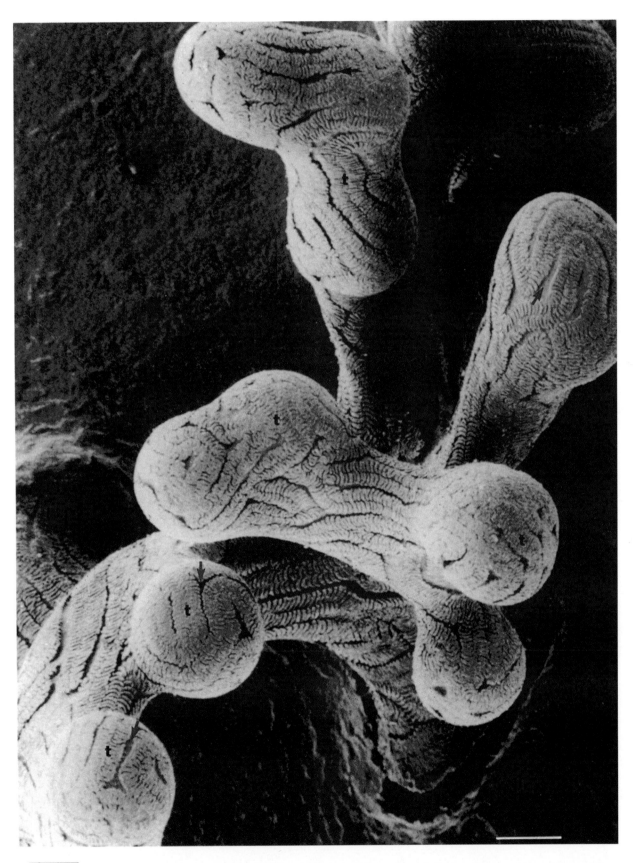

Fig. 45 Scanning electron micrograph of the treelike labyrinthine organ of the bimodal-breather *Clarias mossambicus*, covered by tranverse capillaries (t) separated by tracts (arrows). Scale bar, 0.3 mm.

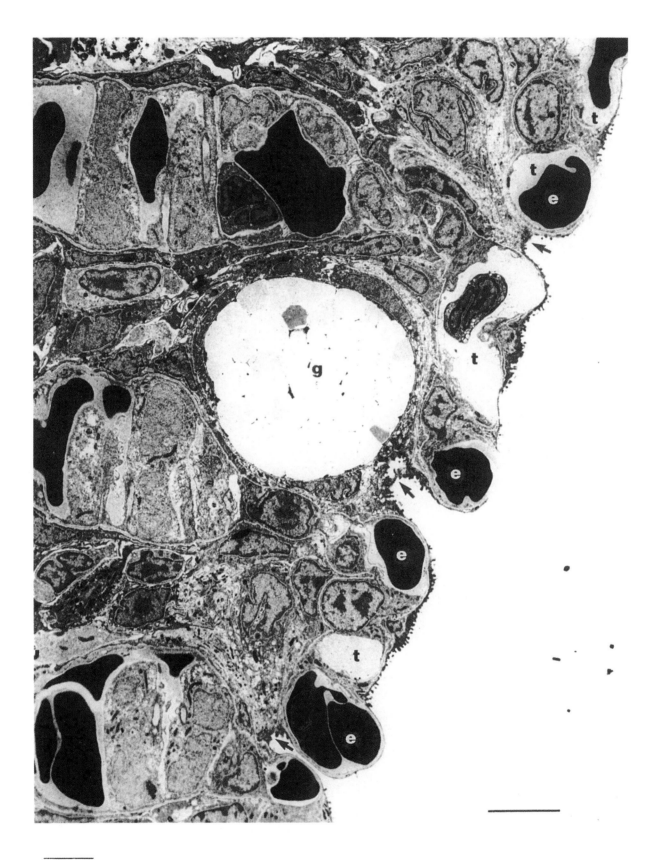

Fig. 46 Transmission electron micrograph of the labyrinthine organ of the bimodal-breather, *Clarias mossambicus* showing tranverse capillaries (t) separated by smooth tracts (arrows) beneath which lie secretory goblet cells (g). e, erythrocytes. Scale bar, 15 μm.

gastrointestinal system are respiratory: residual air is voided through the anus. In the air-breathing fish that use sequestered spaces such as the stomach, intestine, and airbladder, the gills can be simultaneously ventilated: they can take up oxygen while eliminating carbon dioxide, ammonia, and ions. This is not feasible, however, in fish such as the electric eel *Electrophorus electricus* and the knifefish (*Hypopomus*) that use the buccal cavity as a respiratory organ. *Hoplosternum thoracatum* from the Paraguayan swamps (a fish that treks from one pool to another) swallows air continuously: oxygen is absorbed from the posterior part of the intestine that is highly vascularized. The Indian catfish *Clarias batrachus*, living in shallow derelict waters, comes out at night to feed on small animals such as earthworms. During such episodes it uses the suprabranchial chamber membranes and the labyrinthine organs for breathing. Morphologically, the accessory respiratory organs of *C. batrachus* resemble those of *Clarias mossambicus*, an African catfish [Figs. 42-46]. The highly vascularized suprabranchial chamber membrane [Figs. 43, 44] and the treelike (arborescent) labyrinthine organs [Figs. 45, 46] provide highly vascularized, extensive respiratory surfaces. The air-blood barrier in the accessory respiratory organs is much thinner than the water-blood barrier of the gills. Air-breathing fish of India such as the climbing perch, *Anabas testudineus* and the Cuchia eel, *Amphipnous cuchia* are highly terrestrial. They spend much of their time out of water. Specimens of *Anabas* weighing 29 to 51 g acquire 54% of their oxygen needs from air. In *Channa punctatus* and *Anabas testudineus*, the accessory respiratory organs develop during larval and juvenile stages. In *Anabas*, the labyrinthine organs start to develop at the fifth day after hatching but air-breathing does not occur until the 13th or 14th day. The tadpoles do not leave water until development of the air-breathing organs is complete. The secondary reduction of the suprabranchial chambers and the increase in gill respiratory surface area in some species of *Dinotopterus*—Clariidae (of the very deep Lake Malawi of Southern Africa) and that in *Sandelia capensis* (an anabantoid fish in the South African Cape region) points to fish previously well adapted to aerial-breathing reverting to aquatic respiration.

6

Amphibian Lung

1 EVOLUTION OF AMPHIBIANS AND MUTABLE RESPIRATORY MODALITIES

The word amphibian means "double life". The actual evolutionary origin of amphibians is not clearly known. The osteolepiform fish that are thought to have evolved a primal lung or the lungfishes (Dipnoi) are possible candidates that could have given rise to early tetrapods. Although the first vertebrates to invade land some 300 million years ago, due to their strong association with water for important physiological processes such as development, reproduction, and respiration, modern amphibians are a defeated taxon. Their ecological distribution is mainly confined to fresh water, wet, damp, and high rainfall areas—factors that impact greatly on their physiology. Biologically, modern amphibians occupy a pivotal point in the understanding of important evolutionary events such as the change from anamniotic to amniotic eggs, realization of air-breathing, and the transition from aquatic to terrestrial life.

Contemporary amphibians fall into 3 orders. These are: Apoda (= Gymnophiona, caecilians), Anura (= Salentia, frogs and toads), and Urodela (= Caudata, salamanders and newts). The order Apoda ("footless") or Gymnophiona ("naked snake") consists of some 165 species of caecilians. They are highly elusive, vermiform (earthworm-like), tropical, aquatic, semiterrestrial, or fossorial animals. Among amphibians, the caecilians are biologically the least known group. A monophyletic group, they have been isolated from other extant amphibians for at least 70 million years.

Dual subsistence in water and land has obligated development of unique physiological and morphological adaptations in amphibians. Metamorphosis from water-breathing to air-breathing, diversity of habitats occupied, and the multifunctional nature of the lungs (e.g. respiration, vocalization, buoyancy control, and defense), may explain the remarkable structural heterogeneity of the respiratory organs and the multiplicity of the pathways utilized for gas exchange in amphibians. While amphibians generally live in water and moist habitats, a few exceptionally well-adapted species occupy highly desiccating environments, however, with some even living in deserts! By developing an impermeable skin and adopting ureotelism, terrestrial anurans such as *Chiromantis xerampelina* and *C. petersi* endure remarkable degrees of desiccation. They can tolerate water-loss in excess of 60% of their body mass. The highly xerophilic African tree frog, *Chiromantis petersi* leads a characteristically nonamphibious lifestyle. It prefers direct solar insolation and temperatures of 40°C to 42°C. Amazingly, its skin is very well vascularized [Figs. 26, 27].

The need to balance water conservation with gas exchange may explain why there are no large modern amphibians.

2 STRUCTURE AND HETEROGENEITY OF AMPHIBIAN RESPIRATORY ORGANS

During the larval stages of development, amphibians have transient external and internal gills. Subsequently, the lungs and buccopharyngeal cavity become highly vascularized and play important respiratory roles in air. Because it exchanges gases in both water and air, the skin is a dual respiratory organ. Neotenic amphibians retain gills throughout life. Larval caecilians use gills for respiration even while in the egg: the gills are lost soon after hatching. Using gills, the larvae of the viviparous caecilian *Typhlonectes compressicauda* exchange gases and

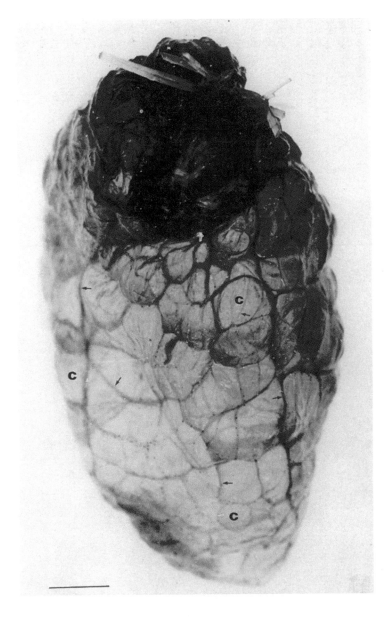

Fig. 47 Photograph of a gross specimen of the lung of the tree frog *Chiromantis petersi*. Arrows, stratified septa that divide the peripheral aspects of the lung into shallow air cells (c). Scale bar, 2.5 mm.

nutrients with the lining of the oviduct. In amphibians, the level of oxygen in the ambient environment mainly regulates the flow of blood to the various respiratory organs, namely the gills, skin, and lungs. Decrease of PO_2 in the lung increases perfusion of the skin and vice versa. Exposed to well-aerated water, the skin receives 20% of the pulmocutaneous blood flow, the quantity decreasing to an insignificant level when the water is hypoxic. In the lungless salamanders (Plethodontidae)—animals that live in cold well-aerated waters—gas exchange occurs exclusively across the skin and the buccal cavity. Agitation of the water during locomotion passively ventilates the skin, increasing its diffusing capacity for oxygen.

Compared with other air-breathing vertebrates, amphibians have some of the simplest lungs. In some species, the internal compartmentalization of the lung consists of internal septa that form one or two levels of air cells, i.e., ediculae [Figs. 47, 48]. The lungs of *Proteus*, *Necturus*, and *Cryptobranchus* are thin-walled, transparent, poorly vascularized, and nonseptate. Such lungs play very little, if any, role in respiration until the midlarval stage. The lungs are mainly involved in hydrostatic regulation, sexual display, and sensory perception. Morphologically and morphometrically, the lungs of anurans and apodans are generally more complex than those of the urodeles. On average, the thickness of the blood-gas barrier in the lungs of Urodela is 2.59 µm, in Apoda 2.35 µm and in Anura 1.89 µm. Some areas of the

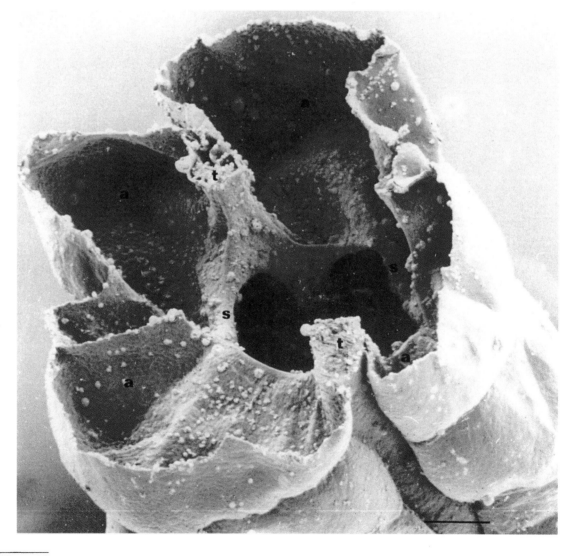

Fig. 48 Photograph of a gross specimen of the tubular lung of the caecilian *Boulengerula taitanus* showing air spaces (a) separated by septa(s). Septa attached to two diametrically placed trabeculae (t). Scale bar, 0. 4 mm.

blood-gas barrier of the lung of the caecilians *Chthonerpoton indistinctum* and *Ichthyophis paucesulcus* are only 1 µm thick; in the tree frog *Hyla arborea* the barrier may even be as thin as 0.6 µm. The lungs of highly terrestrial species, e.g. the toad (*Bufo marinus*) and tree frogs (*Hyra arborea* and *Chiromantis petersi*) [Fig. 47] are particularly elaborate compared to those of caecilians, e.g. *Boulengerula taitanus* [Fig. 48]. The diameters of the air cells range from 1.45 mm in *Rana pipiens* to 2.3 mm in both *Bufo marinus* and *Rana catesbeiana*. Generally, the respiratory surface area in the lungs of the more terrestrial amphibian species is higher than that in lungs of more aquatic ones.

The morphological heterogeneity and multiplicity of the amphibian gas exchangers correspond to the remarkable diversity of habitats and environments they occupy, the lifestyle they pursue, and their pattern of interrupted development. For optimal performance, though amphibians have different respiratory structures, at any one time during their metamorphosis only one structure is best developed. Pulmonary vascularization correlates with the degree of terrestriality, behavior, and tolerance of dehydration. The skin is the primary pathway for gas transfer in the mainly aquatic species while in the more terrestrial ones, it is less prominent or even totally superfluous. *Necturus* held in cool, well-aerated water have better developed gills than animals kept in warmer, poorly aerated water.

"Serpentine" animals (i.e., animals with thin, long cylindrical bodies, e.g. snakes, lizards, and caecilians) typically possess long, tubular lungs. In some species, the left lung is vestigial or totally missing. In aquatic *Typhlonectes compressicauda*, however, as many as 3 lungs develop. The lungs of caecilians are simple, cylindrical, internally subdivided organs [Fig. 48]. A single row of septa, attached to two diametrically located trabeculae, delineate the air cells. The structure of the caecilian lung resembles that of the almost limbless, large, aquatic salamanders such as *Amphiuma* and *Siren*. The fairly elaborate lung of *Boulengerula taitanus* may enable it to procure adequate amounts of oxygen in the hypoxic fossorial environment it inhabits. The hematocrit and the hemoglobin concentration of the aquatic caecilian *Typhlonectes compressicauda* (respectively 38% and 11.3 g. per 100 cm^3 blood) and the terrestrial *B. taitanus* (respectively 40% and 10.3 g. per 100 cm^3 blood) are some of the highest values in amphibians. The large volume of blood (24% to 26% of the body mass) and low P_{50} of the blood hemoglobin (e.g. 3 kPa in *Typhlonectes* and 3.7 kPa in *B. taitanus*) contribute to an efficient respiratory process. While caecilians have a lower body resting metabolic rate than both anurans and urodeles, their aerobic capacity during exercise exceeds that of the latter two groups. Elongation of the lung of caecilians may cause ventilatory limitations during locomotion due to compression of the lung by the trunk muscles. Temporal dissociation between breathing and locomotion has, however, been reported in running lizards, animals with similarly long lungs.

Amphibian lungs are best-developed in Anura where septa intensely subdivide the lung, converting the large central air space into small, stratified air cells. The internal morphology of these relatively elaborate lungs is similar to that of the lungs of dipnoans (lungfishes) [Figs. 40, 41]. The Caudata (e.g. newts) have poorly vascularized lungs with a smooth internal surface. The lung of the characteristically low metabolic rate newt, *Triturus alpestris*, has only 569 capillary meshes per cm^2 while the relatively more metabolically active tree frog *Hyla arborea*, has 652 capillary meshes per cm^2. In *Salamandra, Amphiuma, Megalobatrachus*, and *Siren* (groups that predominantly use lungs for gas exchange), the internal surface of the lung is well subdivided. Their skin is poorly vascularized and the epidermis is very thick (47 µm to 110 µm). In Anura, the skin contributes little in gas exchange. The length of the skin capillaries constitutes only 30% of the total length of the blood capillaries located on the respiratory surfaces. However, adaptively, in two species that live in well-oxygenated high mountain lakes (e.g. *Telmatobius* and *Batrachophrynus*), the lungs are very small, the skin very well vascularized, and the epidermis very thin.

The lungs of most amphibian species, such as *Amphiuma means, Bufo marinus, Boulengerula taitanus* and *Chiromantis petersi*, have a preponderance of smooth muscle

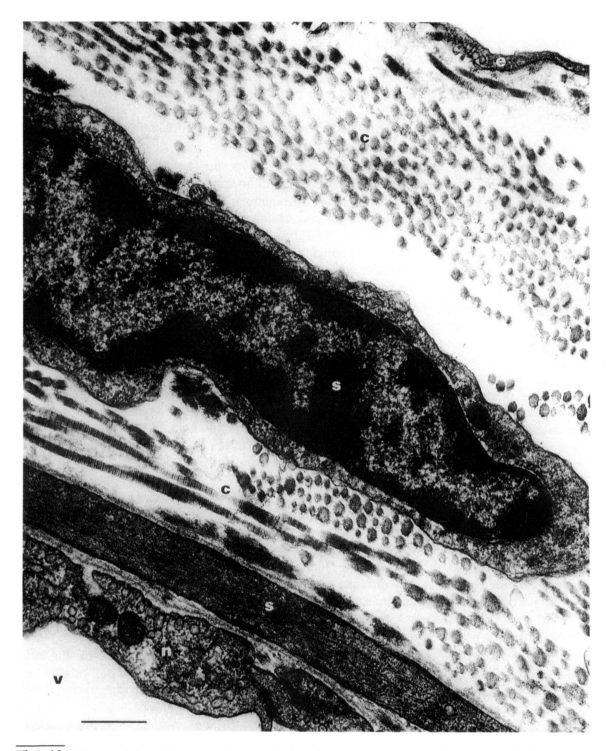

Fig. 49 Transmission electron micrograph showing a cross-section of a septum of the lung of a caecilian, *Boulengerula taitanus*, showing an interstitial space between the epithelial cell (e) and the endothelial cell (n). Interstitial space contains collagen fibers (c) and smooth muscle (s). v, blood vessel. Scale bar, 0.25 μm.

tissue [Fig. 49] that may explain their high compliance. In *Amphiuma*, during expiration the lung virtually collapses, producing an almost 100% turnover of inspired air. The lungs of *Pipa pipa* and *Xenopus laevis* are reinforced with septal cartilages that ensure patency of the air passages. Pulmonary blood capillaries in amphibian lungs [Fig. 50] and those of lungfish [Figs. 51, 52] are exposed to air only on one side. This arrangement is termed the "double

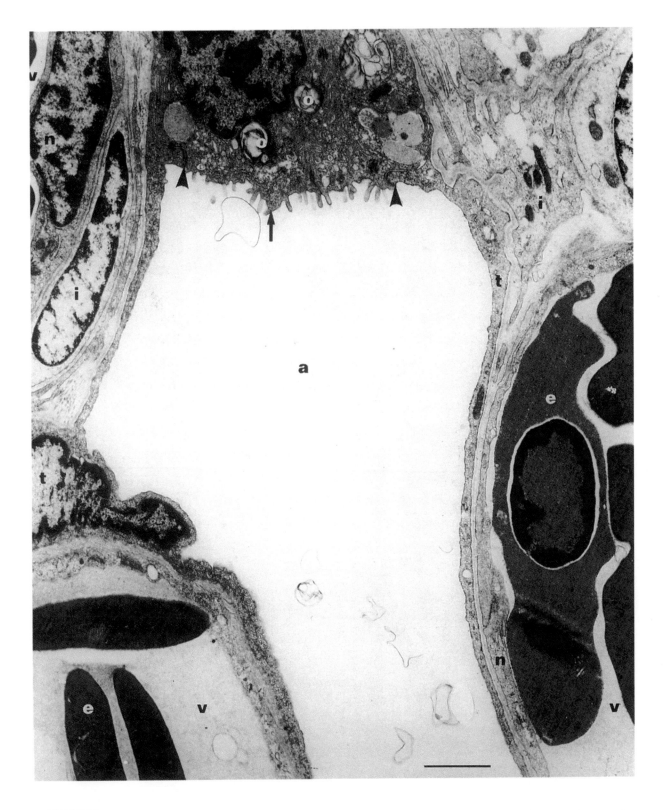

Fig. 50 Transmission electron micrograph of the lung of the tree frog *Chiromantis petersi* showing an air space (a), blood vessels (v), a Type II pneumocyte (arrow) containing lamellated osmiophilic bodies (o) (precursors of the surfactant) and a Type I cell (t) with extensive cytoplasmic extensions. e, erythrocytes; arrowheads, intercellular junctions between Type II and Type I cells; i, interstitial cell; n, endothelial cell. Scale bar, 2.3 µm.

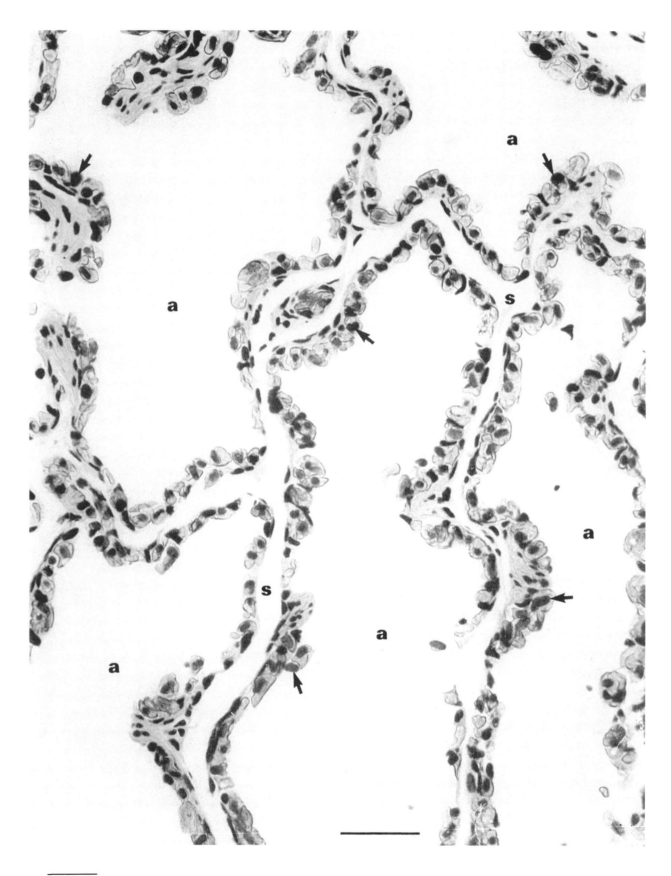

Fig. 51 Light micrograph of the lung of the lungfish *Protopterus aethiopicus* showing air spaces (a) and blood capillaries (arrows) located on opposite sides of the interfaveolar septa (s). The clear spaces in the interfaveolar septa are artifacts of tissue processing. Scale bar, 1 mm.

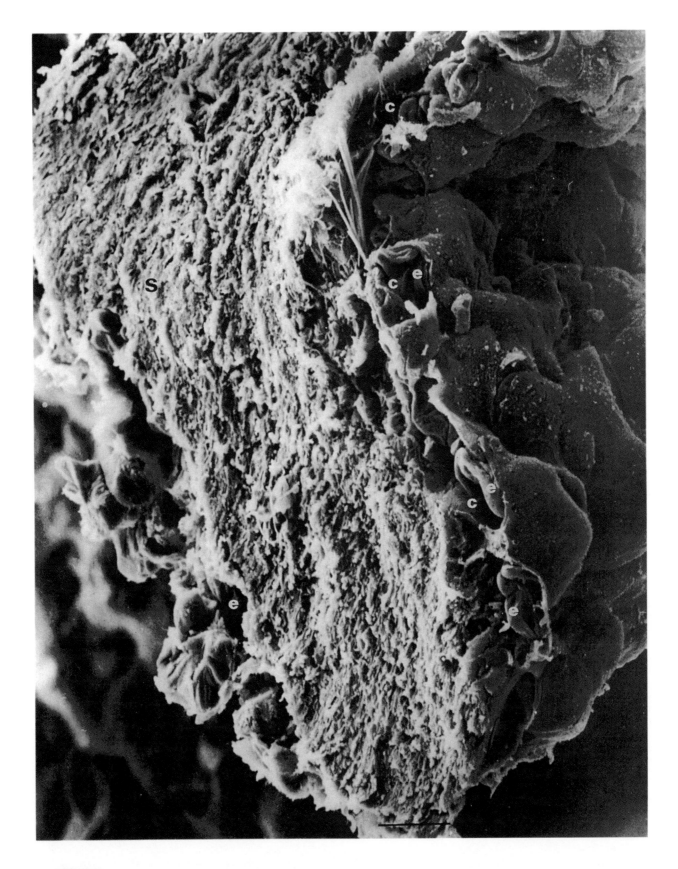

Fig. 52 Scanning electron micrograph of the interfaveolar septum (S) of lung of the lungfish *Protopterus aethiopicus* showing blood capillaries (c) located on opposite sides of the septum. e, erythrocytes. Scale bar, 27 μm.

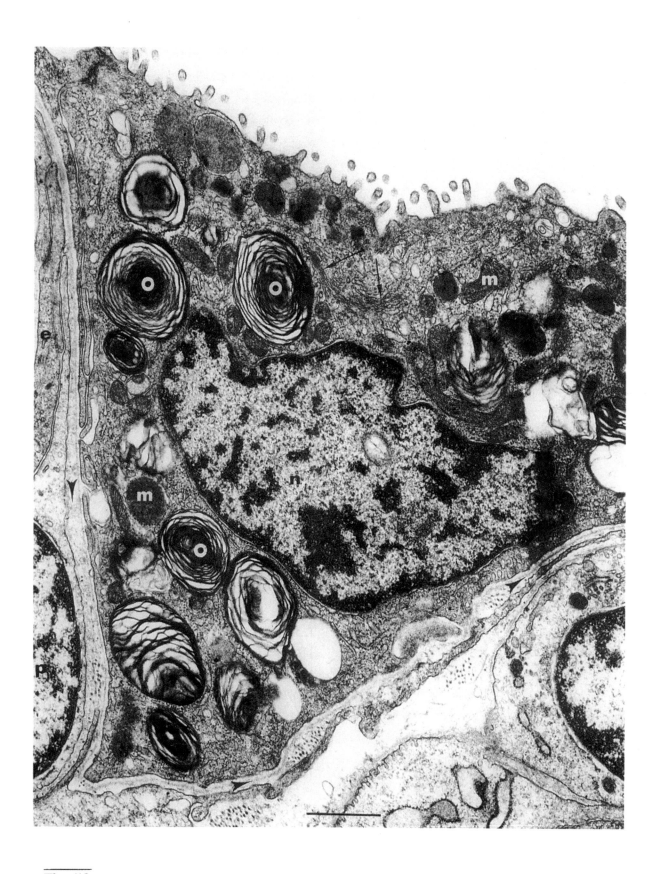

Fig. 53 Transmission electron micrograph of a Type II pneumocyte of the lung of the tree frog *Chiromantis petersi* (boundaries shown by arrowheads). o, osmiophilic lamellated bodies; arrows, Golgi bodies; m, mitochondria; n, nucleus; p, pericyte; e, endothelial cell. Scale bar, 0.7 μm.

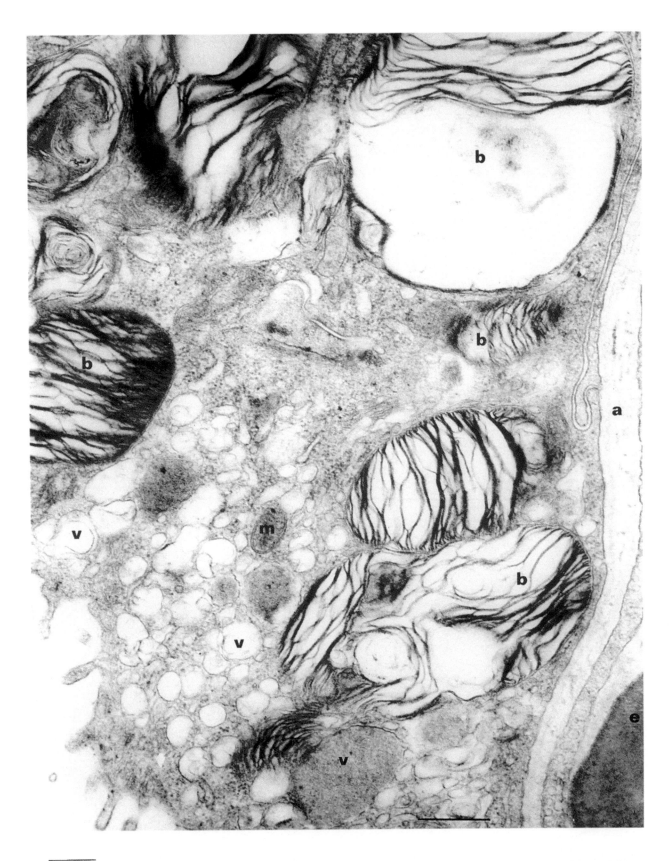

Fig. 54 Transmission electron micrograph of undifferentiated pneumocyte of lung of lungfish *Protopterus aethiopicus* showing surfactant secretory bodies (b) at various stages of development, numerous vesicular bodies (v), and mitochondria (m). a, basement membrane; e, erythrocyte. Scale bar, 0.5 μm.

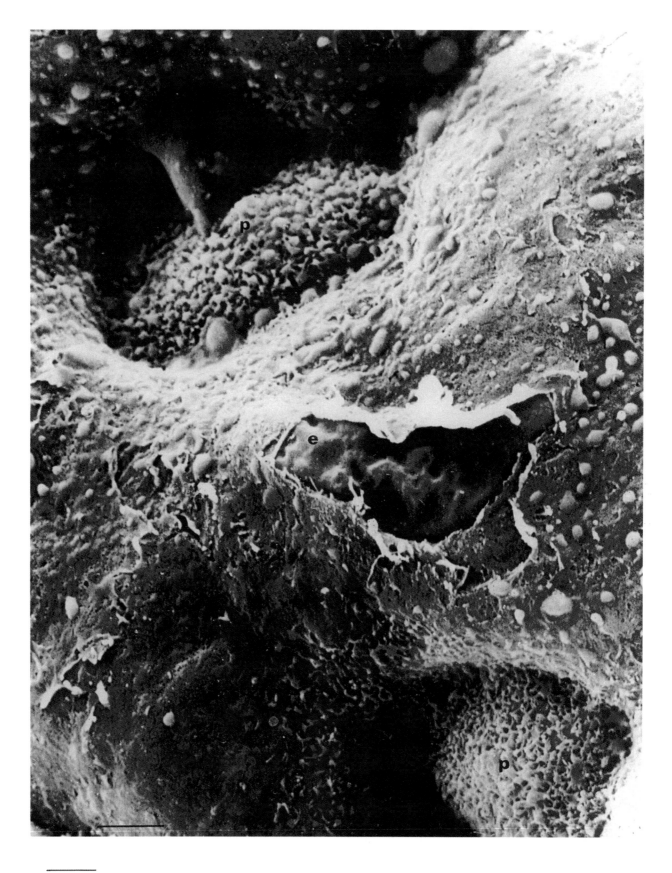

Fig. 55 Scanning electron micrograph of the luminal surface of lung of the lungfish *Protopterus aethiopicus* showing the perikarya of the pneumocytes (p) contained in depressions. Cell perikarya are separated by blood vessels that contain erythrocytes (e). Scale bar, 5 μm.

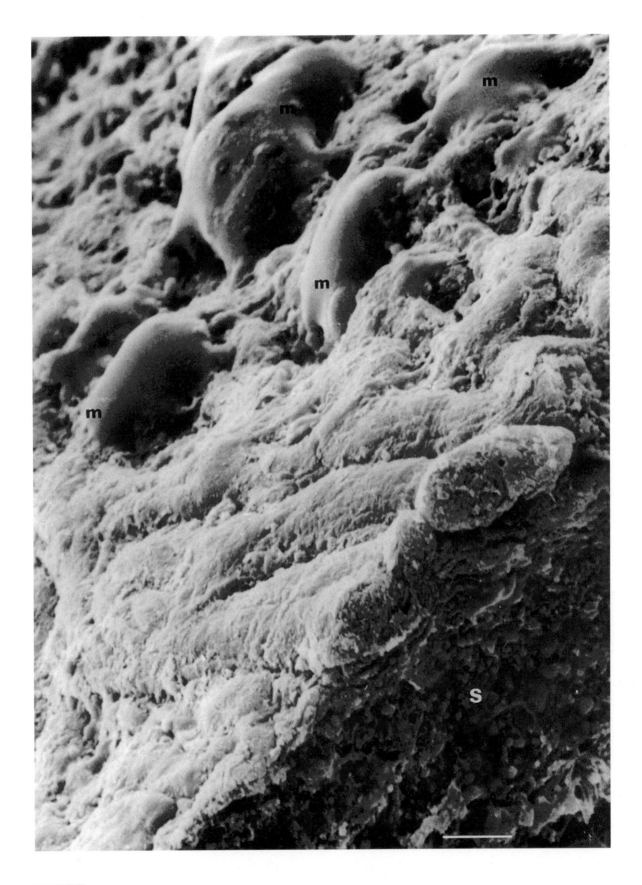

Fig. 56 Scanning electron micrograph of the luminal surface of the lung of the tree frog *Chiromantis petersi* showing macrophages (m). S, septum. Scale bar, 7.6 μm.

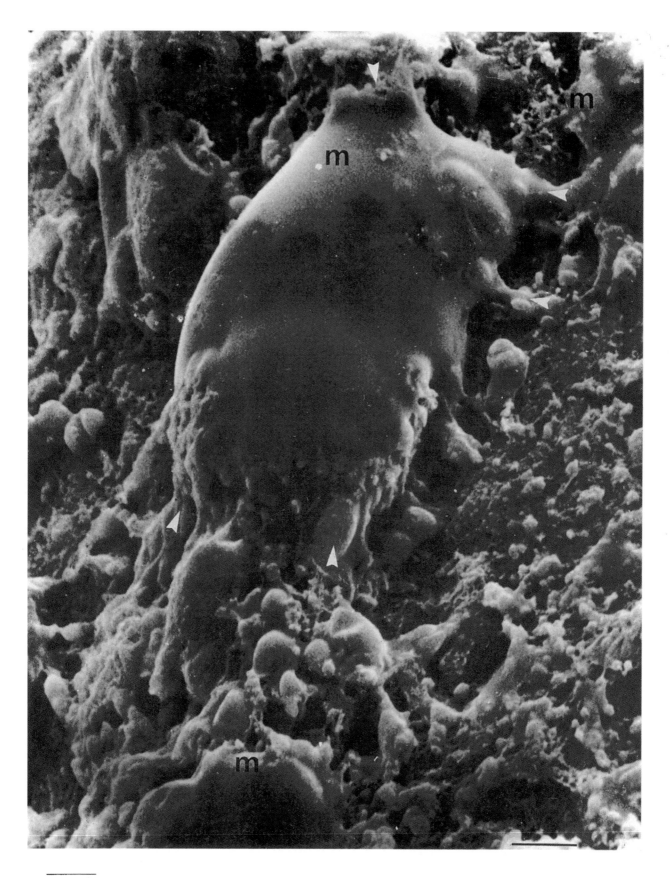

Fig. 57 Scanning electron micrograph showing surface macrophages (m) on lung of the tree frog *Chiromantis petersi*. Arrowheads, filopodia. The cells protect the lungs from pathogenic microorganisms and harmful particulate matter. Scale bar, 6 μm.

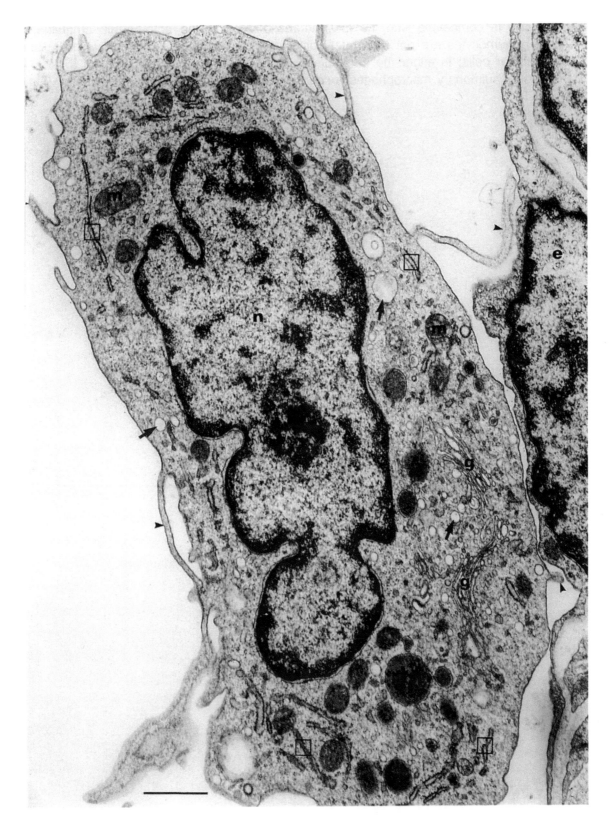

Fig. 58 Transmission electron micrograph showing a surface macrophage on lung of the tree frog *Chiromantis petersi*. Arrowheads, filopodia; squares, rough endoplasmic reticulum; m, mitochondria; arrows, vesicular bodies; n, nucleus; g, Golgi bodies; e, Type II epithelial cell. The cells protect the lungs from pathogenic microorganisms and harmful particulate matter. Scale bar, 0.84 μm.

capillary system". Although this design constrains gas exchange by limiting the respiratory surface area-it is compatible with the low metabolic rates of the ectothermic amphibians. Pulmonary pneumocytes are not completely differentiated into Type I (squamous cells) and Type II (granular cells) in either the amphibian lung [Fig. 53] or those of lungfish [Figs. 54, 55]. Dust cells (pulmonary macrophages) occur on the surface of the amphibian lungs [Figs. 56-58].

7

Reptilian Lung

1 EVOLUTION OF REPTILES, AIR-BREATHING, AND TERRESTRIALITY

Reptiles were the first vertebrates to be adequately adapted for air-breathing and terrestrial habitation. During the about 65 million years of the Carboniferous Period, the climate on Earth was exceptionally warm and moist. Immense forests of primitive trees flourished, particularly around stagnant water masses. The following Permian Period that lasted for about 50 million years was a dry time. With water masses shrinking, it become imperative for animals to lay eggs on land. Only eggs well protected against water loss could survive the conditions. The dry climate saw the downturn of amphibian, a taxon that had retained strong links with water for important physiological processes such as reproduction and respiration. Reptiles had developed an amniotic (cleidoic = "self-supporting") egg and through heavy keratinization of the epidermis a waterproof external cover. With the "sealing-off" of the epidermis, gas exchange was confined to the lungs. Reptiles were the first vertebrates to effectively delink their respiratory physiology and reproduction from water. Aspiratory breathing or costal suctional pumping (pulling air into the lungs), a process effected by contraction of intercostal muscles and the diaphram (when present) first evolved in reptiles. The buccal force pumping mechanism (pushing air into the lungs by compressing the throat) of amphibians was an inefficient method of ventilating the lungs. The diaphragmaticus muscle of the crocodilians pulls the liver and the attached transverse septum backward, enlarging the thoracic cavity, while the obliquus muscle of turtles expands the entire coelom. With gills no longer required, the gill-support skeletal structures of fish were assiged a different role: they served as additional laryngeal, tracheal, and bronchial cartilages that kept the airways patent. As clearly evident from reptilian ascendancy over amphibians, evolution from one major animal group to another entails comprehensive changes of traits that range from structural, physiological, to behavioral ones.

The modern lung with a dual role of oxygen and carbon dioxide exchange as well as acid-base regulation is first encountered in reptiles. The Mesozoic Era that lasted for some 165 million years (from about 230 to about 65 million years ago)—a time when reptiles dominated the terrestrial fauna—is known as "the age of reptiles". Except for those groups that have readapted to water and use the skin for gas exchange, reptiles are exclusively lung breathers. Like amphibians, reptiles display striking pulmonary morphological heterogeneity. To a large extent, the diversity correlates with the diverse habitats occupied and lifestyles pursued. Based on features such as robust pectoral girdle, extensively ossified sternum, and expansive deltopectoral crest of the humerus (features that suggest well-developed flight muscles and possible flight capacity) it has been speculated that the progenitors of birds-the coelurosaurs of the Jurrassic and Cretaceous-had developed complex, multichambered lungs. Moreover, such lungs could have supported some degree of endothermy that is thought to have developed in the Mesozoic and even Paleozoic reptiles. Unfortunately, since soft tissues such as respiratory organs are seldomly preserved, the structure of the lungs of these interesting animals may never be known for certain. The only lungs that have ever been adequately fossilized are those of the Devonian placoderm, *Bothriolepis*, an unfortunately particularly poor representative of vertebrate lungs.

2 QUINTESSENTIAL DESIGN OF REPTILIAN LUNGS

Fundamental differences occur between the structure and function of the respiratory systems of modern reptiles and those of birds [Chapter 8] and mammals [Chapter 9]. These differences particularly pertain to the pattern of air flow and gas exchange efficiency. The reptilian lung constitutes 5% of the body mass. In animals of equivalent body mass, the volume of the lung of a reptile is 7 times larger than that of a mammal. At a body temperature of 20 to 23°C, the diffusing capacity of the reptilian lung is an order of magnitude lower than that of a mammal of the same body mass. This accords with the fact that the aerobic capacity of the ectothermic reptiles is remarkably lower than that of the endothermic mammals. At a temperature of 37°C, a 1 kg lizard consumes 122 ml oxygen per hour, a value only 18% that of an equivalent-sized mammal. The muscle capillary surface area per unit muscle mass of a reptile is about 20% that of a mammal of similar size.

Compared with mammals where endurance exercise induces changes such as increase in tissue oxidative capacities and oxygen consumption, the adaptative response of lizards, e.g. *Amphibolurus nuchalis,* is remarkably different. Trained lizards show reduced heart and muscle masses, but increased liver mass, hematocrit, liver pyruvate kinase, and heart citrate synthetase activities. It has been conjectured that the structural and functional limitations inherent in the design of reptilian lungs, hindered reptiles from attaining endothermic homeothermy.

3 STRUCTURAL HETEROGENEITY OF THE REPTILIAN LUNGS

There is no model reptilian lung. Based largely on the nature and degree of internal subdivision, different morphological groupings of the lungs have been devised. The complex, profusely subdivided multicameral reptilian lungs occur, for example in turtles, monitor lizard, crocodiles, and snakes [Figs. 59-65]. Less elaborate paucicameral lungs occur in, e.g. iguanids, while simple, saccular, smooth-walled unicameral ones are found in, e.g. the teju lizard, *Tupinambis nigropunctatus.* The classification greatly oversimplifies the picture as transitional forms exist. The simplest lungs, e.g. those that occur in Sphenodontia are comparable with the amphibian ones. Such lungs have a central air duct as well as peripherally located, shallow air cells. In less-derived (primitive) species of snakes, e.g. the boas and the pythons, a well-developed left lung as found. Amphisbaenia have atrophied right

---

Fig. 59 Scanning electron micrograph of a preparction of injection of rubber (latex) in the lung (cast) of lung of the monitor lizard *Varanus exanthematicus* showing (ventral aspect) the trachea (t), bronchi (b), and minor lobulations (x). Arrowheads, faveoli. Scale bar, 13 mm.

Fig. 60 Scanning electron micrograph of a preparation of injection of rubber (latex) into the lung (cast) of the chameleon *Chamaeleon chamaeleon* (ventral aspect) showing anterior lobes (x), large air cells (arrowheads), and extensions to the saccular parts of the lungs (arrows). Scale bar, 6 mm.

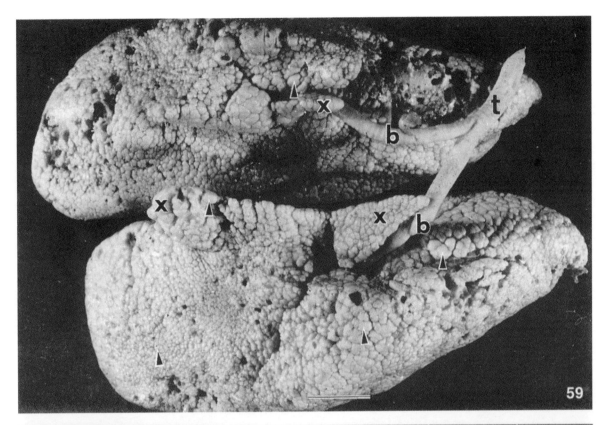

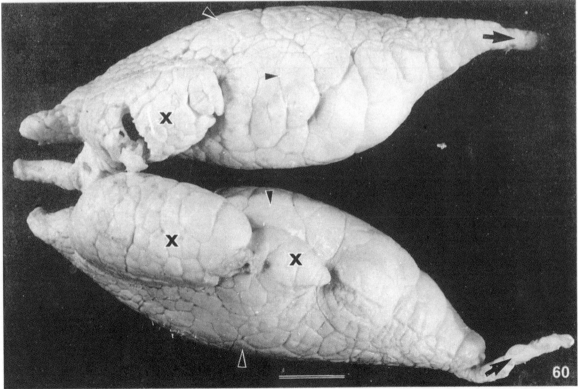

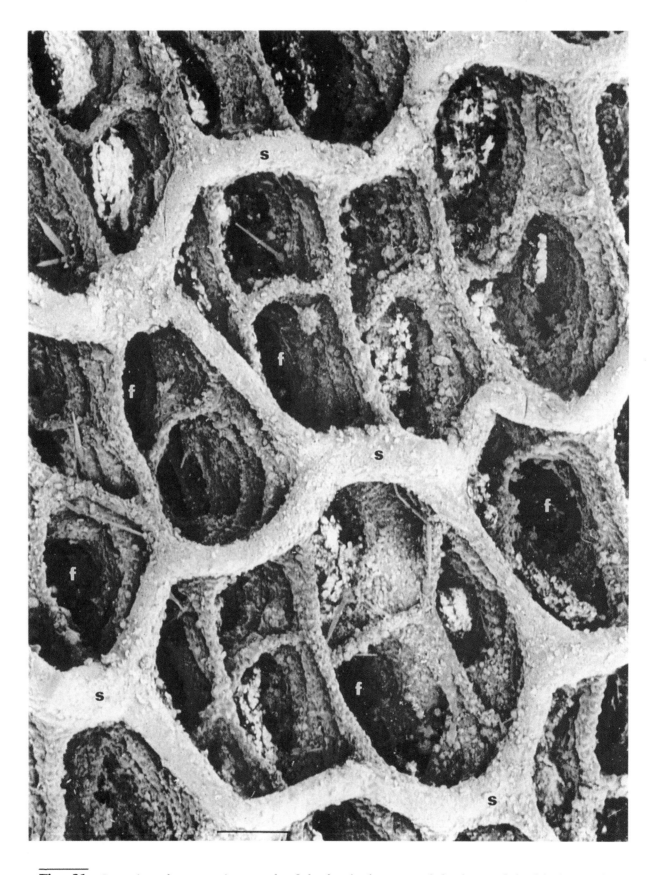

Fig. 61 Scanning electron micrograph of the luminal aspect of the lung of the black mamba *Dendroaspis polylepis* showing major septa (s) that support peripheral partitionings that delineate faveoli (f). Scale bar, 0.1 mm.

Fig. 62 Scanning electron micrograph of lung of the black mamba *Dendroaspis polylepis* showing prominent septa (s) that support the interfaveolar septa (i). Arrowheads, faveolar openings. Scale bar, 0.1 mm.

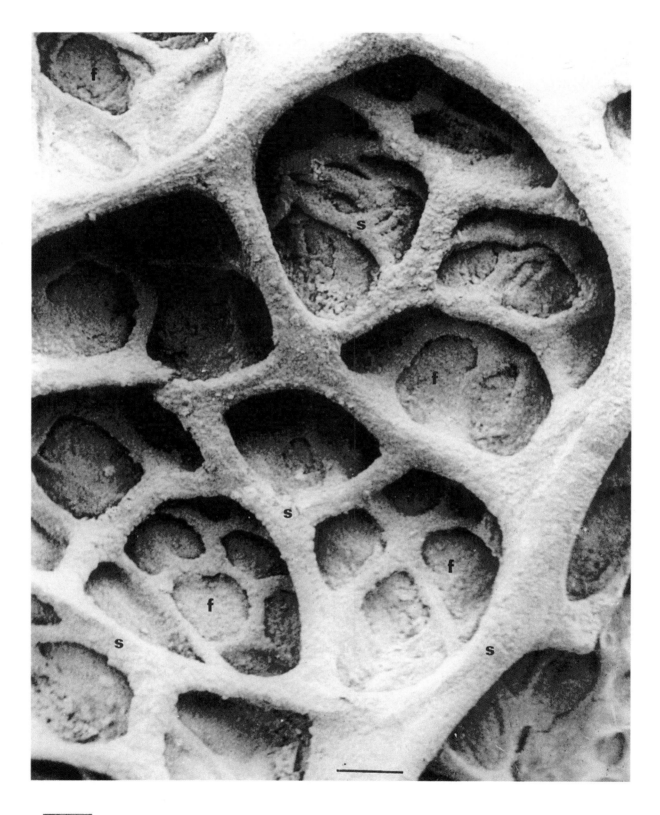

Fig. 63   Scanning electron micrograph of the luminal aspect of lung of the pancake tortoise, *Malacochersus tornieri* showing partitioning of the lung through prominent septa (s) that support smaller peripheral ones which delineate the terminal respiratory air spaces, faveoli (f). Scale bar, 0.25 mm.

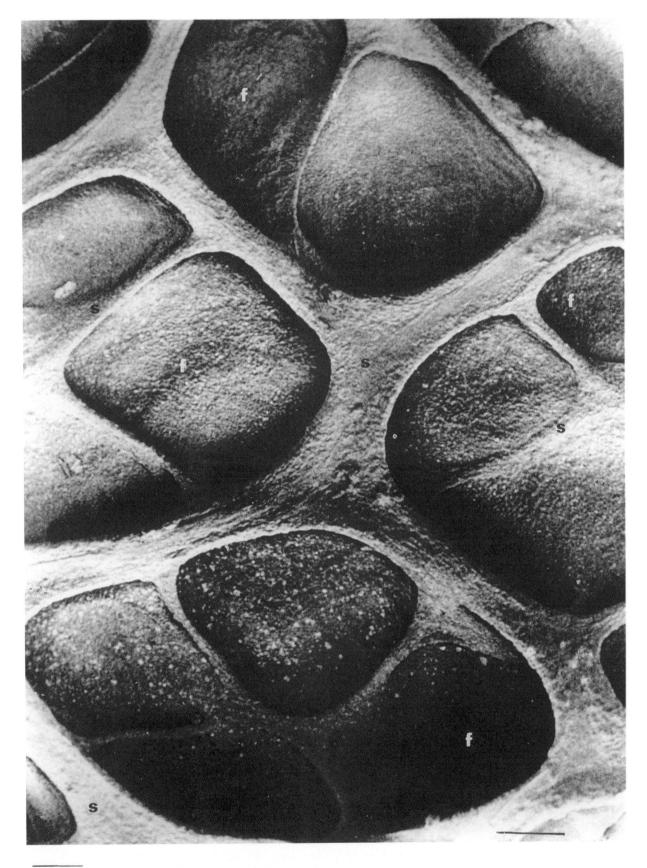

Fig. 64 Scanning electron micrograph of the luminal aspect of internal surface of the lung of the monitor lizard *Varanus exanthematicus* showing respiratory air cells (faveoli) (f) delineated by stratified septa (s). Scale bar, 0.11 mm.

lungs. In the order Squamata, single-chambered lungs preponderate, especially in families Teiidae, Lacertidae, and Gekkonidae. Equally simple lungs occur in family Angioidea. The land-based chelonians have paucicameral lungs, i.e., lungs with two or three levels of peripheral subdivisions: the lungs lack an intrapulmonary bronchus. On the other hand, the marine chelonians have multichambered bronchiolated lungs. Anteroposteriorly, the elongated lungs of snakes and amphisbaenids are divided into two functional zones: the well-vascularized leading part performs the respiratory function while the successive saccular, avascular, and more compliant one serves the roles of storing air and ventilating the anterior gas-exchange part.

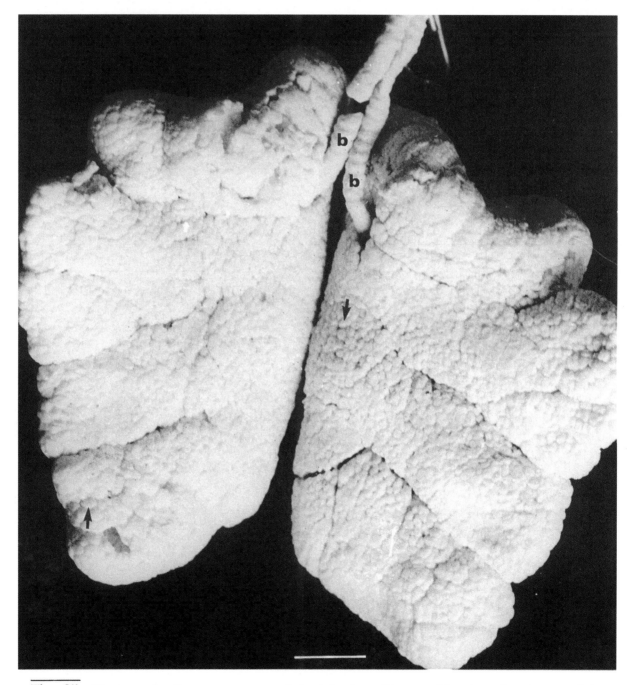

Fig. 65 Photograph of a preparaton made by injection of latex rubber into the airways of the lung of the pancake tortoise *Malacochersus tornieri* showing primary bronchi (b) leading into the lungs. The lungs are poorly lobulated but intensely subdivided into terminal air cells, faveoli (arrows). Scale bar, 1 cm.

Both the bronchoalveolar lung of mammals and the parabronchial one of birds are envisaged to have evolved from the highly complex multicameral lung of reptiles. The lungs of snakes display morphology fairly comparable to that of the lung-air sac system of birds [Chapter 8]: the posterior saccular part stores air and mechanically ventilates the anterior region, in a manner similar to that of the air sacs in the avian respiratory system. Such an arrangement may enhance respiratory efficiency since air passes through the gas-exchange tissue twice, i.e., during inspiration and then during expiration. As the saccular part of the lung is connected in-series with the rest of the lung, the exchange tissue is ventilated tidally. In the avian lung where the air sacs are connected in-parallel to the lung (specifically the primary bronchus versus the paleopulmonic parabronchi-air sac pathways), the exchange tissue is ventilated continuously and unidirectionally. In the crocodilian lung, most of the parenchyma is located in the anterior two-thirds of the lung where blood constitutes 38% to 50% of the total volume. The volume density of the parenchyma in the following reptilian lungs is 25.2% in the tegu lizard, *Tupinambis nigropunctus*, 32.1% in the monitor lizard, *Varanus exanthematicus,* and 25% in the Nile crocodile, *Crocodylus niloticus.* In aquatic species, the reptilian lungs may serve as air stores during apnea, supporting extended dives. In the alligator *Alligator mississippiensis,* the pulmonary oxygen store constitutes 85% of the total lung volume. The faveoli (terminal air cells) make up 40% of the intrapulmonary air in the parenchyma of the crocodilian lung. In the more advanced snakes, e.g. Colubridae, Viperidae, and Elapidae, the left lung is very small and may be lacking in some cases.

Since the lungs of the more primitive species are more homogeneous the morphological heterogeneity in the reptilian lung bespeaks a derived state. Hence, it confers certain functional advantages to the system. The varanids present the highest level of pulmonary development in the suborder Sauria while the pancake tortoise, *Malacochersus tornieri* [Figs. 63, 65] and the monitor lizard, *Varanus exanthematicus* [Figs. 59, 64] have multichambered lungs with branched intrapulmonary bronchi. Internally, the lungs are intensely subdivided. Single-chambered lungs with edicular parenchyma require less energy to ventilate. Such lungs occur in reptiles with low metabolic rates. The trachea of the highly energetic black mamba, *Dendroaspis polylepis* is lined by a ciliated epithelium interpersed with secretory cells [Fig. 66]. The faveolar walls of the black mamba are intensely vascularized [Fig. 67].

Unlike in the lungs of lungfish and most amphibians, the epithelial cells that line the respiratory surface of lungs of certain more advanced reptililes (e.g. crocodiles and varanids) are completely differentiated into Type I and Type II cells. Type III cells (brush cells) and a so-called "mitochondria-rich cell" have been reported in the lung of the turtle *Pseudemys scripta.* Type I cells have remarkably thin cytoplasmic extensions devoid of cell organelles. The much smaller, rather cuboidal surfactant secreting Type II cells are endowed with cell organelles such as mitochondria and Golgi bodies. They are intercalated between Type I cells. Differentiation of the pneumocytes in the lungs of some reptiles and in those of mammals together with birds may be of adaptive significance. By reducing the number and size of the more metabolically active Type II cells, the highly attenuated Type I cells cover most of the respiratory surface. This may enhance respiratory efficiency both by generating a thinner blood-gas barrier and reducing the overall oxygen consumption of the gas exchanger itself. As argued in Chapter 1, an optimal design of a gas exchanger requires that the structure utilizes as little oxygen as possible while transmitting most of it to the blood, the hemoglobin (and via the circulatory system) to the tissue cells. Macrophages (dust cells) occur in the reptilian lungs.

A double capillary system predominates in reptilian lungs [Figs. 68-70]. Smooth muscle tissue preponderates in the interfaveolar septa [Figs. 69-71]. In the tegu and the monitor lizard respectively, smooth muscle tissue constitutes 7.4% and 1.3% of the nontrabecular tissue. The contractile elements of the lung (i.e., smooth muscle and elastic tissue), its saccular nature, and a very efficient pulmonary surfactant determine the overall compliance and hence the ventilatory efficiency of the reptilian lung. Intrapulmonary convective movement

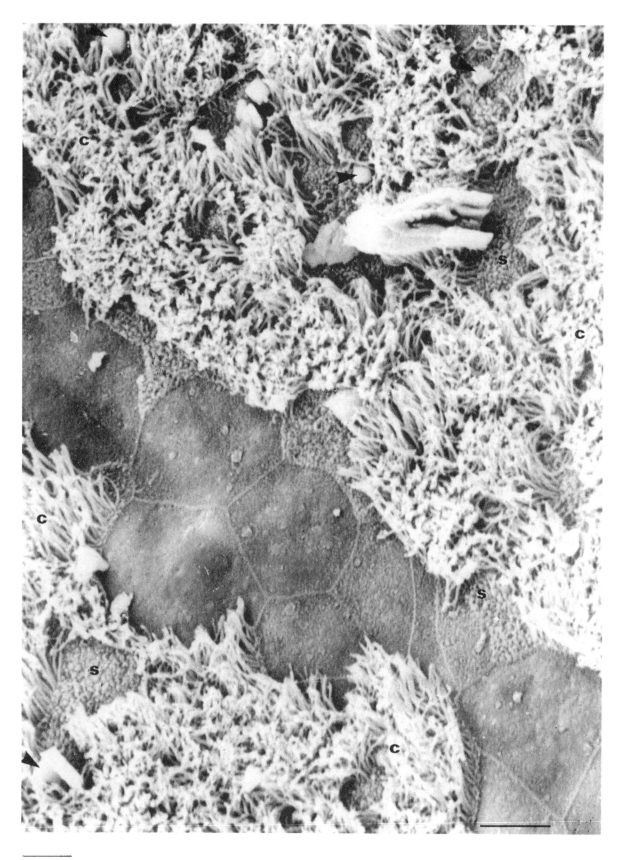

Fig. 66 Scanning electron micrograph of epithelial lining of the trachea of the lung of the black mamba *Dendroaspis polylepis* showing ciliated epithelial cells (c) and secretory cells (s). Arrowheads, secretory material. Scale bar, 10 μm.

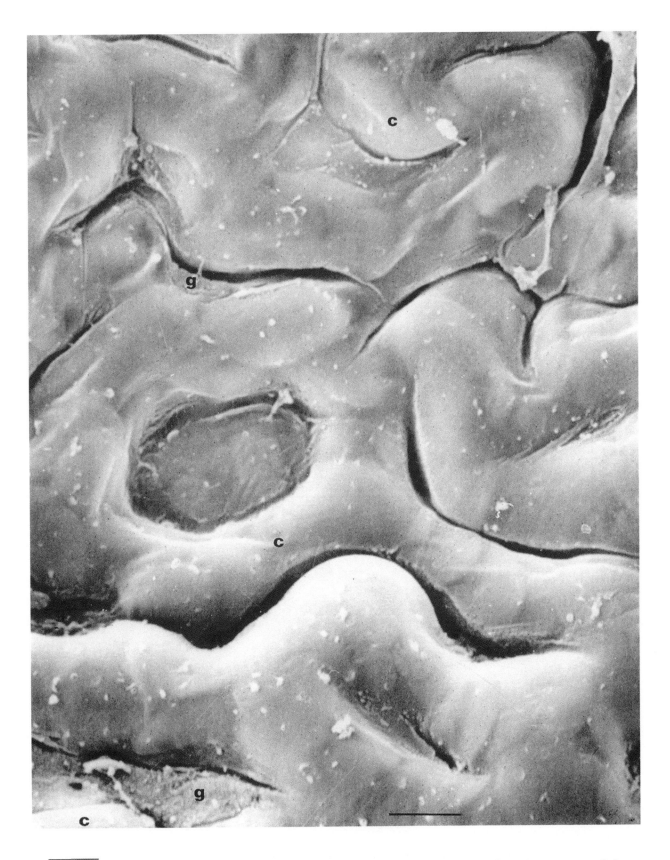

Fig. 67 Scanning electron micrograph of the luminal surface of an interfaveolar septum of the lung of the black mamba *Dendroaspis polylepis* showing blood capillaries (c). g, granular (Type II) pneumocytes. Scale bar, 10 µm.

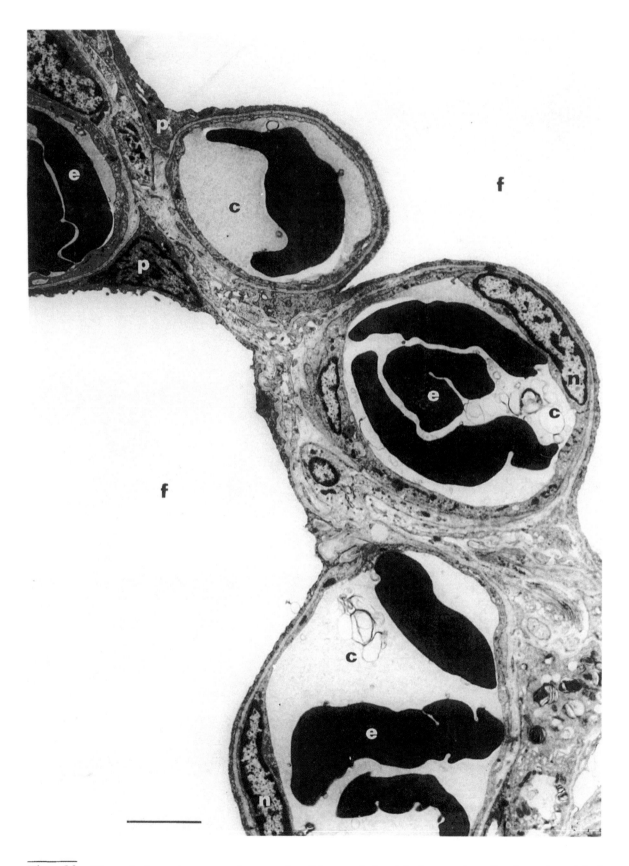

Fig. 68 Tansmission electron micrograph of lung of the black mamba *Dendroaspis polylepis* showing blood capillaries bulging to opposite sides of the air spaces, faveoli (f). c, blood capillaries; n, endothelial cells; e, erythrocytes; p, epithelial cell. Scale bar, 5 μm.

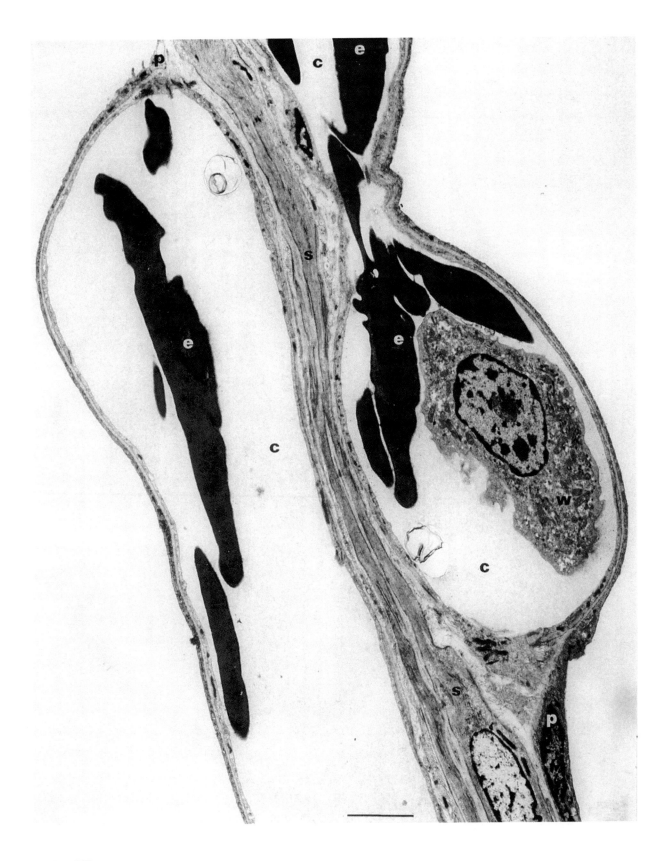

Fig. 69 Transmission electron micrograph showing double capillary arrangement in the lung of the black mamba *Dendroaspis polylepis*. Only one side of the blood capillary is exposed to air. c, blood capillaries; e, erythrocytes; s, interfaveolar septum containing smooth muscle; w, white blood cell; p, epithelial cell. Scale bar, 5 μm.

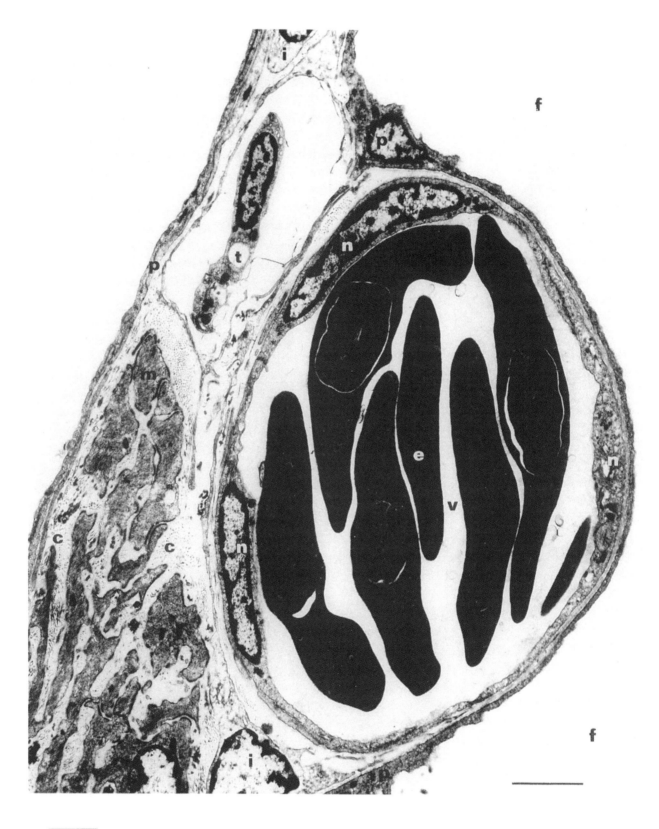

Fig. 70 Transmission electron micrograph of lung of the black mamba *Dendroaspis polylepis* showing a blood vessel (v) bulging into an air space, faveolus (f). Smooth muscle (m) occurs in abundance in the interfaveolar septum. c, collagen; e, erythrocytes; i, interstitial cells; n, endothelial cells; p, epithelial cells; t, interstitial macrophage. Scale bar, 2.5 μm.

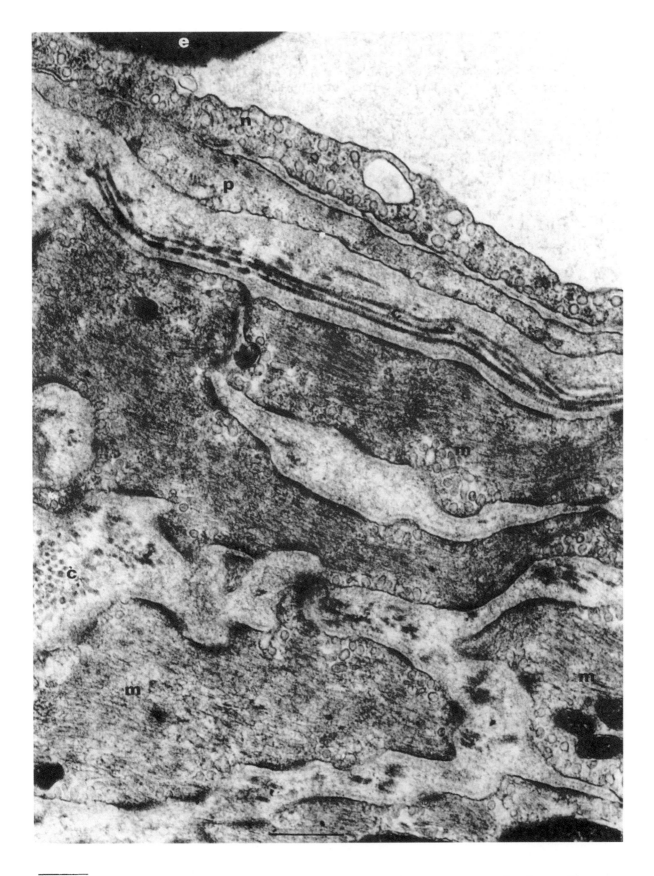

Fig. 71 Transmission electron micrograph showing smooth muscles (m) and collagen fibers (c) in lung of the black mamba *Dendroaspis polylepis.* p, pericyte; n, capillary endothelial cell; e, erythrocyte. Scale bar, 1 μm.

of air has been attributed to contraction of the smooth muscles. Compliance of the lung of the garter snake *Thamnophis sirtalis* of 0.042 (4.2 Pa) ml per cm H_2O per gram is 50 times greater that of the lung of a mouse, a mammal of about the same body mass. At the peak of an expiratory phase, the residual volume of air (18 ml per kg) in the crocodile lung is only 13% of the maximal lung volume. Compliance of the lung of the crocodile of 0.7 (70 Pa) ml per cm H_2O per gram is over 4 times that of the body wall. The volume-specific lung compliance of the multicameral lung of the crocodile is equivalent to that of the much simpler lung of the gecko. This indicates that compliance of the reptilian lung may be an attribute of the parenchymal structure rather than of the type of lung.

8

Avian Lung

1 EVOLUTION OF BIRDS AND THE HIGHLY EFFICIENT PARABRONCHIAL LUNG

Birds evolved from a reptilian stock, coelurosaurs, some 150 million years ago. From fossil records, the crow-sized *Protoavis* and the pigeon-sized A*rcheopteryx lithographica* are the oldest birds. The close phylogenetic affinity between reptiles and birds explains the conspicuous morphological similarities between the two taxa. Birds are frequently referred to as "flying dinosaurs". Flight enabled birds to overcome geographical obstacles and radiate widely into various ecological niches. Among the air-breathing vertebrates, measured in terms of their numerical density, ecological diffusion, and extent of speciation, birds are the most successful taxon. Compared with only 4,200 extant mammalian species, there are about 9,000 avian species. Using molecular genomic techniques, additional species continue to be identified, particularly in certain remote parts of the Earth. Though they now present great diversity in their capacities and modes of locomotion, birds have evolved from a common volant lineage. Some species have, however, totally lost the capacity for flight. Factors such as habitation in predator-free surroundings (e.g. the Australian emu, *Dromiceus novaehollandiae*), attainment of enormous body size (e.g. the ostrich, *Struthio camelus*), and domestication and genetic breeding for fast growth and productivity (e.g. the domestic fowl, *Gallus gallus* variant *domesticus*) are some of the causes for loss of volancy.

After evolving from reptiles, birds attained metabolic capacities between resting and maximal rates of exercise- or cold-induced thermogenesis that were 4 to 15 times higher than those of their progenitors at the same body temperature. Culminating in the establishment of endothermic-homeothermy and adoption of flight, the change should have been closely paralleled by reorganization and refinement of the less efficient reptilian lung. The avian lung-air sac system is considered to have developed from the complex multicameral reptilian lungs like those of the monitor lizard [Figs. 59, 65], turtles [Figs. 63, 65], and snakes [Figs. 61, 62]. Such lungs have a large ventilatory surface area. Chameleon lungs have saccular extensions [Fig. 60] analogous to the air sacs of the avian respiratory system.

2 FLIGHT: A UNIQUE AND ENERGY-COSTLY MODE OF LOCOMOTION

Over the evolutionary continuum, powered flight (defined as the capacity to produce lift, acceleration, and maneuverability at various speeds) has only been attained in four taxa. Chronologically, these include insects (about 350 million years ago), the now extinct *Pterosaurs* (about 200 million year ago), birds (about 150 nillion year ago), and bats (about 50 million year ago). Only a few elite animals have achieved the capacity for volancy. The assortment of animals said to "fly", e.g. the freshwater butterfly-fish (*Pantodon buchholzii*) of West African rivers, the parachuting frog of Borneo (*Rhacophorus dulitensis*), the flying snakes of the jungles of Borneo (*Chrysopelea* sp.), the flying squirrel of North America (*Glaucomys volans*), the flying lemur (*Cyanocephalus volans*), and the East Indian gliding lizard (*Draco volans*) are strictly acrobatic passive gliders or parachutists that utilize part of their body to slow down a fall by using drag and lift. They have not had to grapple with the inordinate aerodynamic and aerobic demands that have challenged active flyers such as insects, birds, and bats. The fact that only four phylogenetically distant taxa (among many) have ever attained volancy attests to the extreme selective pressure that animals must endure to attain this novel mode of locomotion.

Energetically, flight is highly costly. A significant metabolic barrier discriminates the volant vertebrates from the nonvolant. In flight, an animal consumes oxygen at a rate 2 to 3 times that of a ground-dwelling one at maximum exercise. The oxygen consumption of a running pigeon is 27.4 ml per minute and that in flight (at a speed of 10 meters per second) 77.8 ml per minute. Flying in turbulent air or during ascent, a bird increases its oxygen consumption for short intervals by a factor of 20 to 30 times. A champion human athlete can achieve such increases for only a few minutes. Though highly energy-expensive in absolute terms, powered flight is a very efficient form of locomotion. At fast speeds, the energy expended per unit distance covered is less than that incurred in most other forms of active physical displacement. For example, in the bats *Phyllostomus hastatus* and *Pteropus gouldii*, the energy needed to cover a certain distance is respectively only one-sixth and one-fourth that required by the same size nonflying mammal. At their most optimal speeds, the minimum energy cost of flying over a given distance for a 380 gram bird is about 30% that of a 380 gram mammalian runner.

Regarding aspects such as speeds attained, endurance, and altitude at which some species perform, avian flight is remarkable. The peregrine falcon, *Falco peregrinus*, has been reported to dive on its prey at speeds in excess of 180 kph; in its annual migration, the Arctic tern, *Sterna paradisea,* flies from pole to pole, a distance of 35,000 km; and collision between a Ruppell's griffon vulture, *Gyps rueppellii,* and a jet plane at an altitude of 11 km has been recorded. At that altitude, the barometric pressure is about 24 kPa (i.e., 20% that at sea level), the PO_2 in the inspired air less than 5.3 kPa (closer to 2.7 kPa if hyperventilation brings the PCO_2 to about 0.67 kPa), and the ambient temperature about -60°C. The capacity of birds to survive, let alone fly, under such extreme conditions is unequalled among other animals. It requires exceptionally efficient lungs.

3 STRUCTURE OF THE AVIAN RESPIRATORY SYSTEM

Among the air-breathing vertebrates, the lung-air sac system of birds is structurally the most complex and functionally the most efficient respiratory organ. Its design is so elaborate that it is impossible to correctly discern the airflow pathways in the system by mere physical examination of the lung. While in the amphibian, the reptilian, and mammalian lungs the respiratory and the compliant elements are integrated, in birds, the lung (gas exchanger) has been divorced from the air sacs (ventilatory structures) [Plate V]. The avian lung is compact and virtually rigid [Plate V; Figs. 72, 73]. Deeply attached to the ribs, its volume changes only by a mere 1.4% between the respiratory cycles. Compression of the lung does not cause

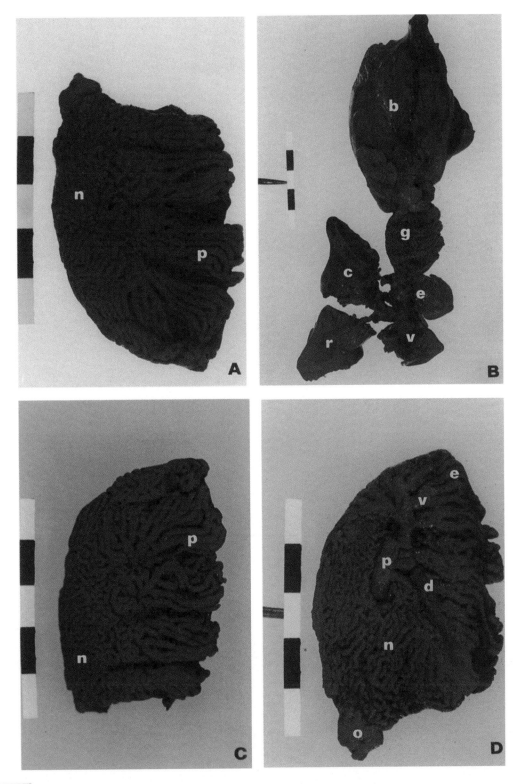

Plate V Photograph of lung and air sacs of the domestic fowl, *Gallus gallus* variant domesticus and the muscovy duck *Cairina moschata* prepared by injection with plastic fluid (setting) material. A: Lateral view of lung of the domestic fowl showing paleopulmonic (p) and neopulmonic parabronchi (n). B: Lung-air sac system of the domestic fowl showing lateral view of the lung (g) intercalated between air sacs, namely abdominal (b), caudal thoracic (c), cranial thoracic (r), cervical (e), and interclavicular (v) air sacs. The air sacs have been physically separated for clarity. C: Lateral view of lung of the muscovy duck showing paleopulmonic (p) and neopulmonic (n) parabronchi. D: Medial view of lung of the domestic fowl showing a primary bronchus (p), medioventral secondary bronchi (v), and mediodorsal secondary bronchi (d). e, paleopulmonic parabronchi; n, neopulmonic parabronchi; o, ostium opening into the abdominal air sac. Scale bars, 1 cm.

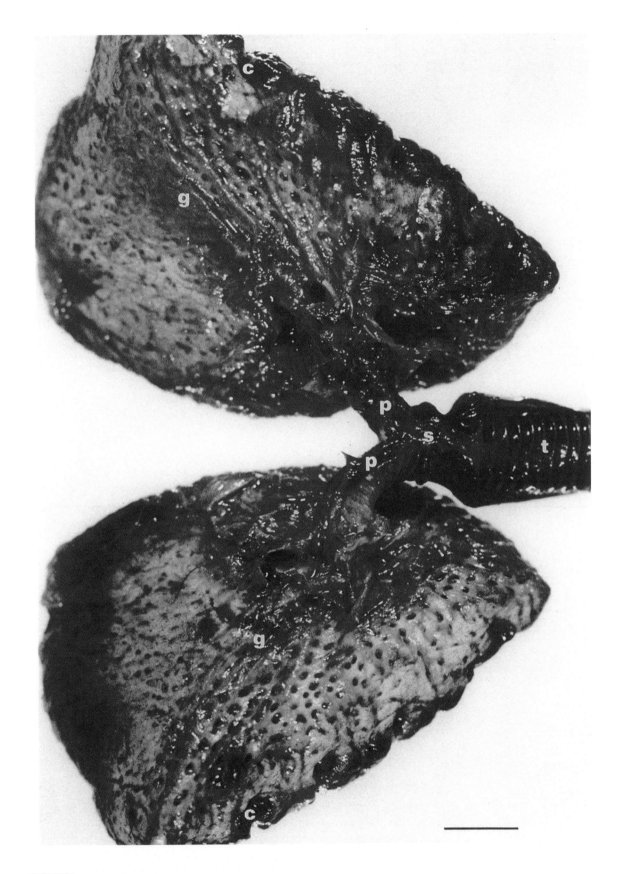

Fig. 72 Photograph of a gross specimen of the lungs (g) of the domestic fowl *Gallus gallus* variant *domesticus* showing the trachea (t), syrinx (s), and extrapulmonary primary bronchi (p). c, costal sulci. Scale bar, 1 mm.

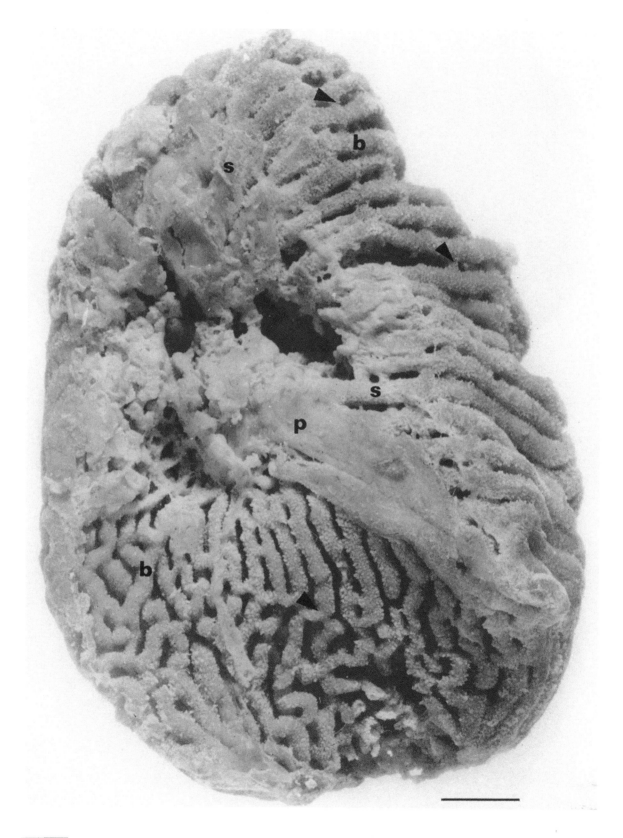

Fig. 73 Photomicrograph of a preparaction of injection of rubber (latex) into the airways of lung of the domestic fowl *Gallus gallus* variant *domesticus* showing the primary bronchus (p), secondary bronchi (s), and parabronchi (b). Arrowheads, parabronchial anastomoses. Scale bar, 5 mm.

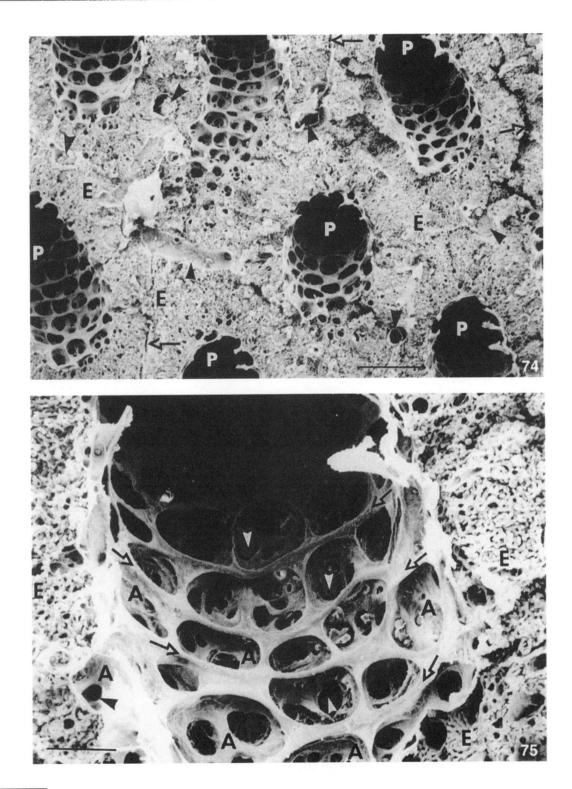

Fig. 74 Scanning electron micrograph of parabronchi of lung of the domestic fowl *Gallus gallus* variant domesticus showing parabronchial lumina (P) surrounded by exchange tissue (E). Arrows, interparabronchial septa; arrowheads, interparabronchial blood vessels. Scale bar, 70 μm.

Fig. 75 Scanning electron micrograph of parabronchus of lung of the domestic fowl, *Gallus gallus* variant domesticus showing atria (A) that arise from the parabronchial lumen. The atria open into the infundibulae (arrowheads). Arrows, atrial muscles; E, exchange tissue. Scale bar, 25 μm.

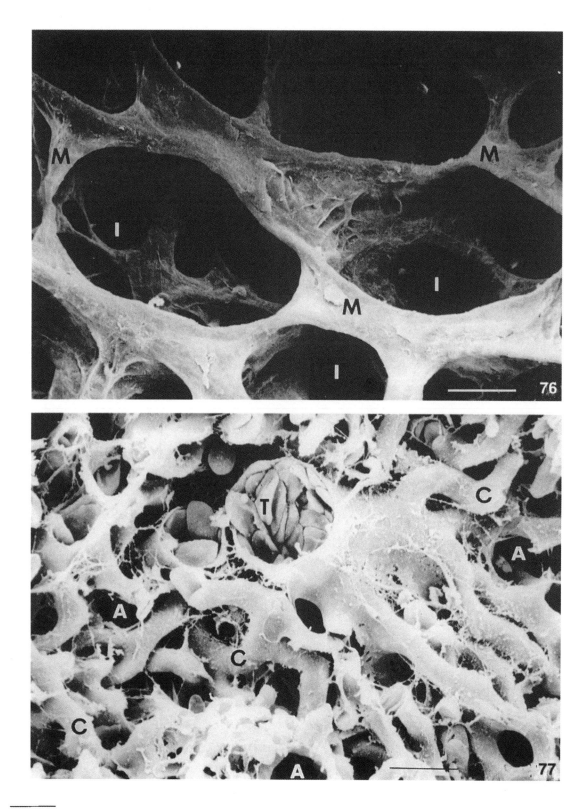

Fig. 76 Scanning electron micrograph of the atria of lung of the domestic fowl, *Gallus gallus* variant *domesticus* showing atrial muscles (M) that give rise to infundibulae (I). Scale bar, 16 µm.

Fig. 77 Scanning electron micrograph of exchange tissue of lung of the domestic fowl *Gallus gallus* variant *domesticus* showing an intraparabronchial blood vessel (T) giving rise to blood capillaries (C). The blood capillaries interdigitate closely with the air capillaries (A). Scale bar, 2.5 µm.

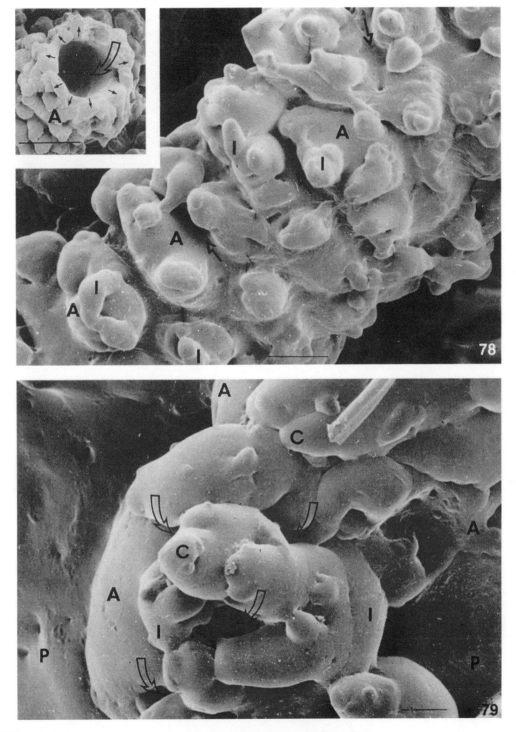

Fig. 78 Scanning electron micrograph of a preparation made by injection of latex rubber into the airways of lung of the domestic fowl *Gallus gallus* variant *domesticus* showing atria (A) extending from its luminal aspect and giving rise to infundibulae (I). Arrows, interatrial septa. Scale bar, 0.2 mm. Insert shows the parabronchial lumen (open arrow) and peripheral extensions (small arrows) of atria (A) from the parabronchial lumen. Scale bar, 0.5 mm.

Fig. 79 Scanning electron micrograph of a preparation made by injection of latex rubber into the airways of a bird lung showing atria (A) arising from parabronchus (P) of the domestic fowl *Gallus gallus* variant *domesticus*. The atria give rise to infundibulae (I) that terminate in air capillaries (C). The air capillaries interdigitate with blood capillaries (spaces shown with open arrows). Scale bar, 0.5 mm.

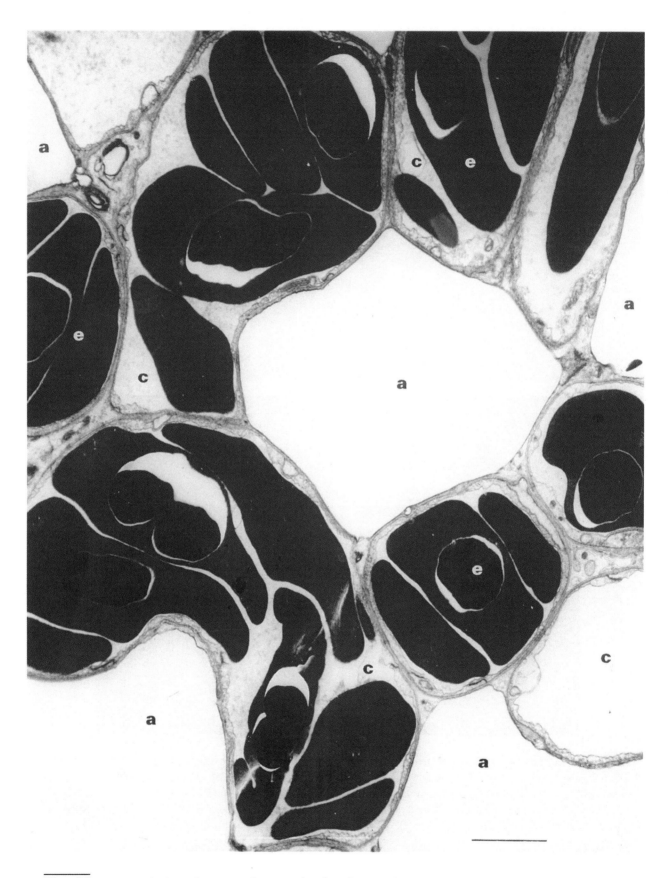

Fig. 80 Transmission electron micrograph of exchange tissue of lung of the black-headed gull *Larus ridibundus* showing air capillaries (a) and blood capillaries (c) that contain erythrocytes (e). Scale bar, 3.5 μm.

significant collapse of the air capillaries! The avian lung is ventilated continuously and unidirectionally (caudocranially) by synchronized action of capacious, avascular air sacs. Functionally, two sets of air sacs, namely cranial and caudal groups, exist. The cranial air sacs comprise the cervical and the interclavicular air sacs while the caudal group consists of the abdominal, cranial thoracic-, and caudal-thoracic air sacs [Plate V].

Though compared with a mammal of similar body mass a bird has a smaller lung per unit body mass by a factor of 27%, the respiratory surface area (i.e, the surface area of the blood-gas barrier) is 15% greater in the lung of a bird. This is explained by the fact that the

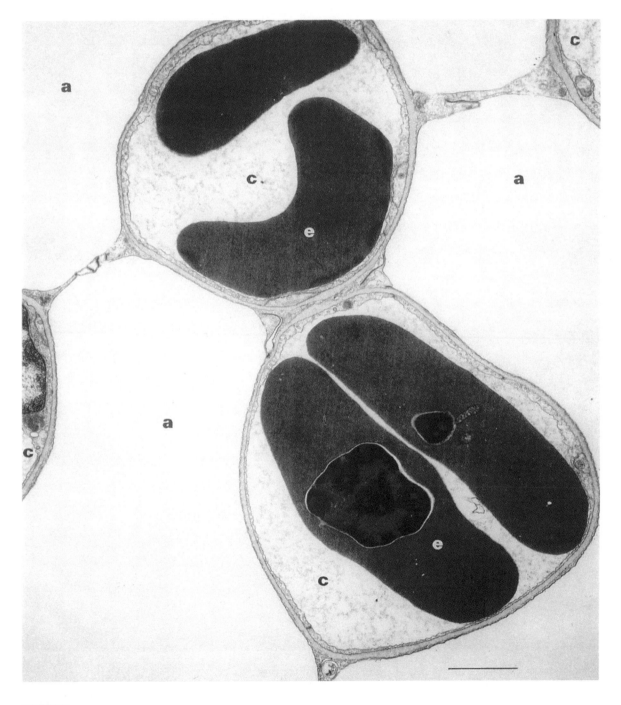

Fig. 81 Transmission electron micrograph of exchange tissue of lung of the domestic fowl *Gallus gallus* variant *domesticus* showing blood capillaries (c) containing erythrocytes (e). The blood is exposed to air virtually all around. a, air capillaries. Scale bar, 2.8 μm.

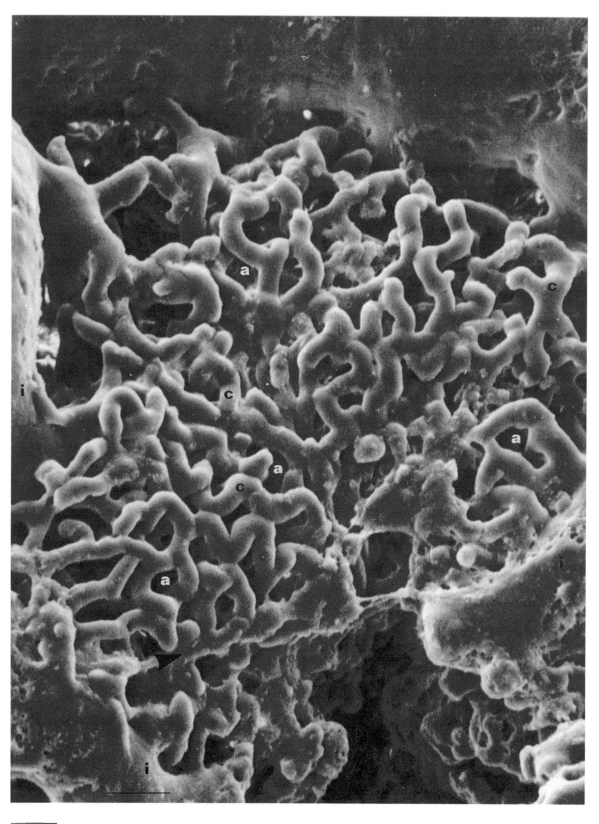

Fig. 82 Scanning electron micrograph of a preparation made by injection of latex rubber into the vascular system of a bird lung showing a blood capillary system (c) of lung of the domestic fowl *Gallus gallus* variant domesticus. i, intraparabronchial arterioles; arrowhead, interatrial septum. Unfilled spaces (a) constitute air capillaries that interdigitate with the blood capillaries. Scale bar, 15 μm.

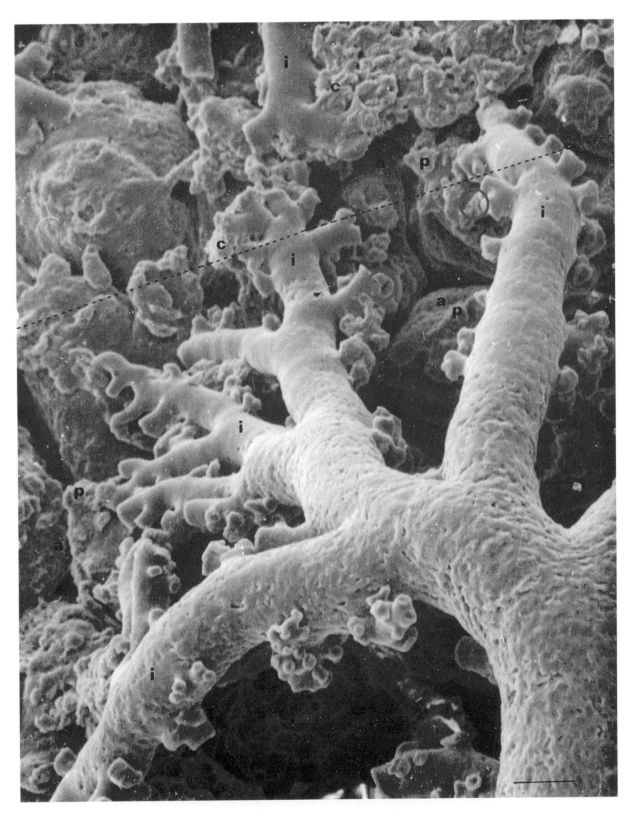

Fig. 83 Scanning electron micrograph of a preparation made by injection of latex rubber into the vascular and airway systems of lung of the domestic fowl *Gallus gallus* variant *domesticus* showing the cross-current relationship between the blood flow in the intraparabronchial arterioles (i) and the airflow in the parabronchial lumen (orientation shown by dashes). Circles, points where air- and blood capillaries interact; a, atria; c, blood capillaries; p, air capillaries. Scale bar, 40 μm.

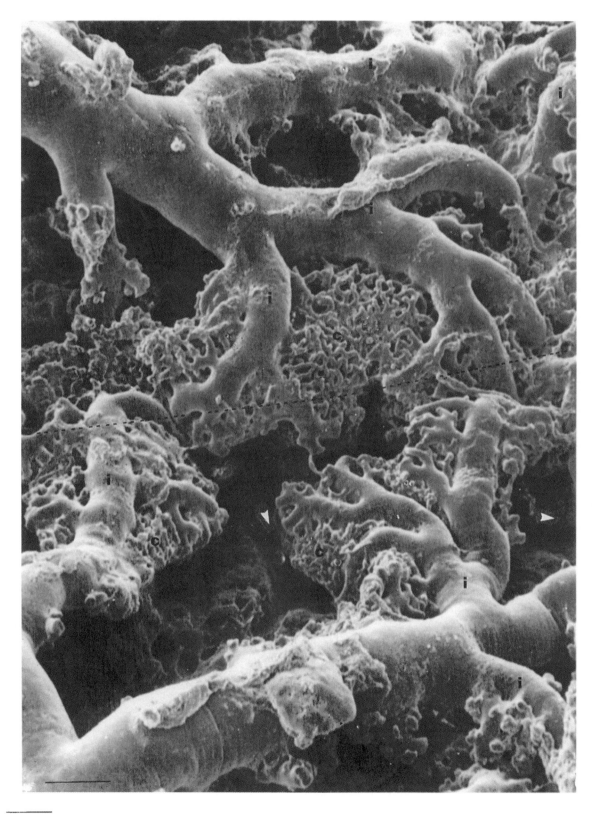

Fig. 84 Scanning electron micrograph of a preparation made by injection of latex rubber into the vascular and airway systems of lung of the domestic fowl *Gallus gallus* variant *domesticus* showing the cross-current relationship between the blood flow in the intraparabronchial arterioles (i) and airflow in the parabronchial lumen (orientation shown by dashed line). Blood capillaries (c) cover and interdigitate with air capillaries emanating from the infundibulae and atria. Arrowheads, interatrial septa. Scale bar, 20 μm.

parenchyma (parabronchial gas-exchange tissue) of the avian lung has been intensely compartmentalized. The diameters of the air capillaries (the terminal gas-exchange components) range from 8 μm to 20 μm. The extreme subdivision of the exchange tissue was granted by the rigidity of the lung, after relegation of the ventilatory compliance to the air sacs. In a nonexpansile lung, surface tension was no longer a limiting factor in setting the ultimate size of the terminal respiratory units. While in mammals compliance of the respiratory system (excluding the thoracic walls) is determined by the terminal parts of the tracheobronchial system (mainly the alveolar spaces), in the avian respiratory system compliance is totally confined to the air sacs. Futhermore, the rigidity of the avian lung may have allowed development of a thin blood-gas barrier.

For such a highly diversified taxon, morphologically, the avian respiratory system is strikingly homogeneous. Subtle differences exist, however. The most important pertain to the development and arrangement of the parabronchi (tertiary bronchi) [Figs. 74, 75]. Two sets, the paleopulmonic and neopulmonic parabronchi occur. The paleopulmonary parabronchi are located on the dorsocranial region of the lung [Plate V] and constitute about two-thirds of the volume of the lung. The neopulmonic set is located ventrocaudally and comprises about one-third of the lung volume. The paleopulmonic parabronchi are arranged in parallel stacks while the neopulmonic ones are irregularly arranged and anastomose more profusely [Fig. 73; Plate V]. While the airflow in the paleopulmo is continuous and unidirectional (in a caudocranial direction), that in the neopulmo is tidal, i.e., changes in direction with the phase of breathing. The paleopulmonic parabronchi develop before the neopulmonic ones. Only paleopulmonic parabronchi occur in the lungs of primitive birds such as the kiwi and the penguin. Reaching maximum development in the small highly evolved passerine species, the neopulmo is variably developed in the more conserved species. The overall evolutionary and functional significance of the paleo- and the neopulmo systems is unclear. There are no morphometric differences in the sizes of the air- and blood capillaries in the two sets of parabronchi. However, tidal ventilation of the neopulmo may confer a site for carbon dioxide recycling. This may be important in averting respiratory alkalosis from excessive carbon dioxide washout across the undirectionally and continuously ventilated paleopulmonic parabronchi, especially when an animal is panting under thermal stress. The ostrich *Struthio camelus* for example, can pant continuously for as long as 8 hours without experiencing acid-base imbalance.

4 QUANTITATIVE MORPHOLOGY OF THE AVIAN LUNG

The entire volume of the respiratory system in birds (i.e., volume of the lungs, air sacs, and pneumatic spaces) constitutes about 20% of the total body volume, with the value in the mute swan *Cygnus olor*, for instance, being as high as 34%. The volume of the lung and the air sacs in a bird is 3 to 5 times that of a mammalian lung and 2 times that of a reptile of equivalent body mass. The total volume of blood in the avian lung may constitute as much as 36% of the volume of the lung, with 58% to 80% of this blood being in the capillaries. The capillary blood volume in the avian lung is 2.5 to 3 times that in the mammalian lung, where only 20% occurs in the alveolar capillaries. The gas-exchange tissue (the parenchyma) of the avian lung [Figs. 74, 75] forms about 46% of the lung volume: in the mammalian lung, the parenchymal volume density is about 90%. The surface density of the blood-gas barrier, i.e., the surface area per unit volume of the parenchyma, ranges from 172 $mm^2.mm^{-3}$ in the domestic fowl *Gallus gallus* variant *domesticus* to 389 $mm^2.mm^{-3}$ in the hummingbird *Colibri coruscans*. On average, the values in birds are 10 times greater than those of mammals. It is important to reemphasize at this point that in the avian lung, an extensive respiratory surface area has been shrewdly generated under constraints of a much smaller lung and parenchymal volume. This has occurred through extreme subdivision of the parabronchial gas-exchange tissue, a process that has only been allowed by the presence of a compact lung. The compartmentalization of the gas exchange tissue is so intense that the epithelial surface area of the air capillaries essentially equals that of the capillary endothelium.

From the parabronchial lumen of the avian lung, the atria give rise to as many as 8 infundibulae that in turn give rise to air capillaries [Figs. 74-81]. The blood capillaries arise from the intraparabronchial arterioles that in turn derive from the interparabronchial arteries [Figs. 82, 83]. Topologically, the air and blood capillaries closely interdigitate with each other [Figs. 82, 83]. Capillary loading (i.e., the ratio of pulmonary capillary blood volume to respiratory surface area)—a parameter that denotes the degree of exposure of blood to air in a gas exchanger—is close to unity. While a double capillary system occurs in the lungs of lungfish [Fig. 50], amphibians [Figs. 51, 52], and reptiles [Figs. 69, 68] and a single capillary exists in the adult mammalian lung [Chapter 9], a "diffuse capillary system" (formed by profuse intertwining of the air- and blood capillaries) occurs in the avian lung. Conceptualized from a three-dimensional perspective, in the avian lung the pulmonary capillary blood is virtually suspended in air [Figs. 80, 81], providing excellent exposure of blood to air. This contributes greatly to the gas-exchange efficiency of the lung.

5 FUNCTIONAL DESIGN OF THE AVIAN LUNG

In the avian lung, the geometric arrangement between the flow of air in the parabronchial lumen and the centripetal flow of venous blood from the periphery of a parabronchus [Figs. 83, 84] is described by a crosscurrent model. Previously, from the unequivocally high efficiency of the avian lung rather than from any empirical evidence, it was assumed that a countercurrent system occurred in the parabronchial lung. Reversal of the flow of one of the respiratory media in a countercurrent system converts the system to a concurrent one [Figs. 6, 8, 9]. This drastically reduces the gas-exchange efficiency. In a crosscurrent system, however, such procedure only changes the arterialization sequence of the blood in the blood capillaries. Overall respiratory efficiency is not affected [Figs. 9, 10b, 83, 84].

All factors equal, the gas-exchange efficiency of a countercurrent system should exceed that of a crosscurrent one. In the parabronchial lung, however, a multicapillary serial arterialization system is inbuilt in the crosscurrent arrangement. Venous blood of uniform composition is delivered to all parts of the parabronchi through infinitely numerous blood capillary segments that arise from intraparabronchial arterioles [Figs. 83, 84]. This prolongs the interaction between air and blood along unidirectionally and continuously ventilated parabronchi (paleopulmonic), considerably enhancing respiratory efficiency. The overall quantity of oxygen in the arterial blood returning to the heart from the lung (after oxygenation in the lung) essentially ensues from an additive, i.e., a "pooling up effect," of minute quantities of oxygen acquired at infinitely many points along the parabronchial exchange tissue [Figs. 83, 84]. The efficiency of the avian lung is demonstrated by the fact that under certain conditions, e.g. hypoxia and exercise, the PO_2 in the arterial blood may exceed that in the end-expired air. In the evolved gas exchangers, the PO_2 in the arterial blood can only exceed that in the respiratory fluid medium leaving the gas exchanger only in the countercurrent system of the fish gills and the crosscurrent one of the avian lung [Fig. 9].

6 STRUCTURAL-FUNCTIONAL CORRELATIONS IN THE DESIGN OF THE AVIAN LUNG

Lungs of the small energetic species of birds show distinct pulmonary morphometric specializations. The small passerine species that have a high metabolic rate and operate at a higher body temperature (42°C) have superior lungs. Compared to those of the nonpasserine species, their lungs have an extensive respiratory surface area, a thin blood-gas (tissue) barrier, and a high pulmonary capillary blood volume. Gliding and soaring birds such as seagulls that expend less energy in flight have relatively lower values compared with those that have to flap their wings continuously while in flight. Among birds, hummingbirds have a high hemoglobin concentration in blood, a high oxygen-carrying capacity of blood, erythrocyte counts, and large hearts. By developing very small body size, hummingbirds have occupied an ecological niche earlier only inhabited by insects. Regarding energetic cost for life, hummingbirds pay a great price. Over a 24-hour cycle, they literally have to eat continuously and in some cases even estivate to conserve energy. Hummingbirds have a heart rate as high as 1,300 times per minute (during hovering flight), a wing beat frequency of up to 80 times per second, a heart size about twice that of most other birds, and whole body circulatory time of about 1 second. During hovering, they consume oxygen at a rate of 40 ml per gram per hour. The mitochondrial volume density of the flight muscles (the volume fraction of the muscle fiber occupied by mitochondria) is as high as 35%.

The highest mass-specific respiratory surface area of 87 cm^2 per gram has been reported in the small and highly energetic violet-eared hummingbird, *Colibri coruscans,* and the African rock martin, *Hirundo fuligula.* The value is substantially greater than that of 43 cm^2 per gram in the shrew *Crocidura flavescens,* one of the smallest and highly energetic mammal. Extremely thin blood-gas barriers (harmonic mean thickness) of 0.090 μm and 0.099 μm respectively occur in the rock martin and the violet-eared hummingbird. The thickness of the blood-gas barrier in the shrew (0.334 μm) is three times thicker than the thinnest barriers in the avian lung. The flightless galliform species (e.g. the domestic fowl and the guinea fowl, *Numida meleagris*) have relatively low pulmonary diffusing capacities compared to the more energetic birds. The lowest pulmonary morphometric diffusing capacity in birds occurs in the emu, *Dromiceus novaehollandiae,* a large flightless Australian bird that in its natural setting encounters few predators. The morphometric parameters of the lung of the ostrich *Struthio camelus,* a large bird exposed to numerous predators in the African savanna, are close to those of the volant species. The humboldt penguin, *Spheniscus humboldti,* an efficient diver, has an exceptionally thick blood-gas barrier of the lung (0.530 μm). The thick barrier may grant capacity for withstanding high hydrostatic pressures during dives.

9

Mammalian Lung

1 EVOLUTION OF MAMMALS AND THE BRONCHOALVEOLAR LUNG

The start of the Tertiary period saw mammals overtake and eclipse reptiles as the dominant terrestrial vertebrate group. The advent of placental mammals (subclass: Eutheria) was the pinnacle of advancement of the taxon. Mammals may have survived the dinousaur demise of the end-Cretaceous (after the two groups had coexisted for over 100 million years) mainly because of their relatively small size, equivalent to that of a rat or a mouse or smaller. This would have allowed them to fit into more hospitable habitats inaccessible to the larger dinosaurs. The succcession of dinosaurs by mammals was thus "a victory by default": mammals were in the right place at the right time and were well prepared when disaster struck! Like mammals, birds endured the pressure very well: at least 22 avian lineages predate the Cretaceous-Tertiary catastrophe.

Among vertebrates, the mammalian lung has been best studied structurally and functionally. The human lung in particular, without valid justification, is assumed to be a model of the vertebrate lungs. Moreover, unwaranted extrapolations regarding their morphology and physiology are made to other gas exchangers. Given the abundance of literature on the mammalian lung and the need for brevity in this book, only general aspects concerning its structure are presented. However, structural-functional considerations of the lung of bats are covered in some detail in the light of recent interesting data. The question: "why are bats the only mammals that fly?" appears in many popular magazines, in professional journals as well as in zoological and physiological books. For earlier lack of comprehensive data, unsatisfactory answers have often been given.

Compared with the avian lung [Chapter 8], the mammalian lung developed much earlier from the complex multicameral reptilian lung [Chapter 7]. Dichotomous branching of the airway and vascular system [Plate VI; Fig. 85] provides an extensive respiratory surface area while occasioning approximation of the respiratory media, blood and air. The blind-ending [Figs. 5, 6, 9, 10], tidally ventilated mammalian lung falls far short of the level of respiratory efficiency attained by the continuously, unidirectionally ventilated avian one [Chapter 8], the only other taxon that has achieved endothermic-homeothermy. It is notable that birds operate at a relatively much higher body temperature ($40^{o}C$-$42^{o}C$) compared to mammals ($38^{o}C$).

Among the various mammalian species (except for differences such as bifurcation of the bronchial tree, lobulation, and topographic relationships between the airway and the vascular

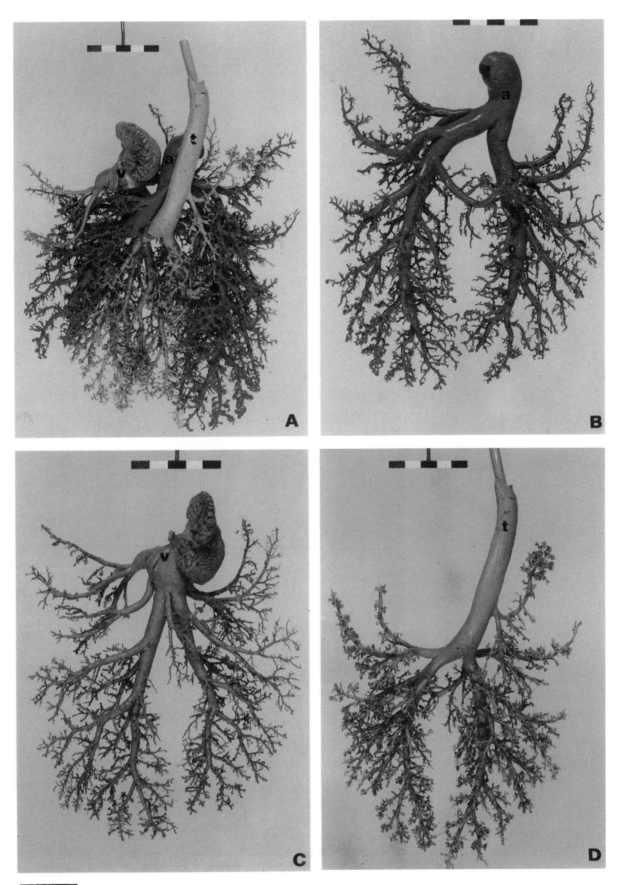

Plate VI Photograph of a cast of tracheobronchial (t), arterial (a), and venous (v) systems of the lung of the pig *Sus scrofa* prepared by injection of plastic fluid (setting) material. A: The three systems branch dichotomously, but closely interact. B: Arterial system; C: Venous system; D: Tracheobronchial system. Scale bars, 1 cm.

systems), the mammalian lung is structurally remarkably homogeneous. However, a noteworthy exception occurs in the the naked mole rat (*Heterocephalus glaber*), a small, eusocial, poikilothermal, fossorial, bathyergid rodent that lives in thermostable burrows of the arid semideserts of the horn of Africa: it retains neotenic lungs throughout life. Even during its remarkable longevity for a rodent of its size, alveolation of the parenchyma is never completed. Large terminal air spaces, in some cases lined by columnar to cuboidal epithelium, occur in some parts of the adult lung. From morphological and morphometric perspectives, the lungs of the nonhuman primates are more appropriate models for the study of the human pulmonary structure and function. Structural-functional correlations occur in design of the mammalian lung. Small, highly metabolically active species have more refined lungs. The diffusing capacity of the lung of the Japanese waltzing mouse (*Mus wagneri*), a pathologically hyperactive animal, is 55% greater than that of a normal mouse. Shrews, the smallest extant mammals with the highest resting metabolic rates, have exceptionally specialized lungs with an extensive respiratory surface area and a particularly thin blood-gas barrier. Because it is intensely subdivided, the parenchyma of the lung of the shrew has been described as "honeycombed". Lungs of high-altitude mammals have greater diffusing capacities for oxygen than those of sea-level dwellers. Lobectomy induces compensatory growth, with the remaining lobes soon attaining the initial diffusing capacity for oxygen.

The bronchial system of the mammalian lung branches progressively [Plate VI; Fig. 85], terminating in the rather hexagonal air cells termed the alveoli [Figs. 85-88]. The large airways are lined by a pseudostratified, ciliated, columnar epithelium with goblet cells [Figs. 89-92]. With bronchial bifurcation, the elaborate epithelium lining the airways gradually changes to simple columnar, simple cuboidal, and simple squamous states. The mucus-secreting goblet cells disappear ahead of the ciliated cells. Mucous lining is important for humidification of inhaled air, protection of the underlying epithelial cells as well as trapping and clearance of particles by a mucous escalator system. The alveolar surface is mainly lined by completely differentiated Type I and Type II cells [Figs. 93, 94]. The former are thin expansive cells devoid of organelles while the later are small, rather cuboidal cells endowed with various organelles. The Type II cells [Fig. 95] secrete surfactant, phospholipid material that reduces surface tension at the air-water interface. Free macrophages (phagocytes) [Fig. 96] occur in varying numbers on the alveolar surface. Filopodia and numerous lysosomal bodies characterize the cells. They protect the lung by phagocytosing pathogenic microorganisms. On the thick (supporting) part of the interalveolar septum [Fig. 97], a thin epithelial cell, an interstitial space (containing connective tissue elements such as collagen and elastic tissue), and an endothelial cell occur. The thin (respiratory) side of the interalveolar septum (the blood-gas barrier) consists of an epithelial cell, a common basement membrane, and an endothelial cell.

2 STRUCTURE AND FUNCTION OF THE BAT LUNG

Among mammals, bats (order: Chiroptera) are unique from their capacity of flight. As in birds, volancy allowed bats to overcome geographical obstacles, granting them unprecedented adaptive radiation and speciation. Consisting of about 800 species, one in five mammalian species is a bat! After the human being, *Myotis* (family: Vespertilionidae) is the most widely dispersed mammalian genus on Earth. However, notwithstanding their numerical density and ecological distribution, due largely to their elusive nocturnal lifestyle (a niche to which they were permanently relegated by the more successful birds that evolved flight some 50 million years earlier), bats remain animals of curiosity, myth, and unfortunately prejudice. The phylogenetic affinity between bats and the mainstream mammals is unclear. Scandentia (Tupaiidae or tree shrews), primates, and Dermoptera are possible candidates of taxa from which bats may have arisen. Comprising suborders Megachiroptera and Microchiroptera, bats are considered by some authorities to be monophyletic while others deem them to be diphyletic. The earliest reliably known fossil remains of a bat are those of *Icaronycteris index*

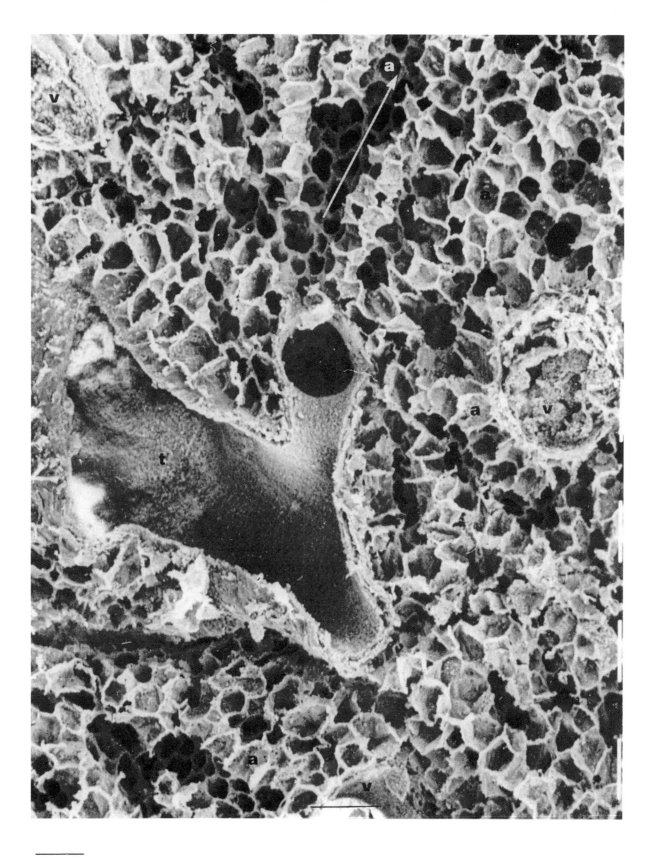

Fig. 85 Scanning electron micrograph of lung of the bat *Tadarida mops* showing a terminal bronchus (t) giving rise to a respiratory bronchiole (arrow) that gives rise to alveoli (a). v, blood vessels. Scale bar, 87 μm.

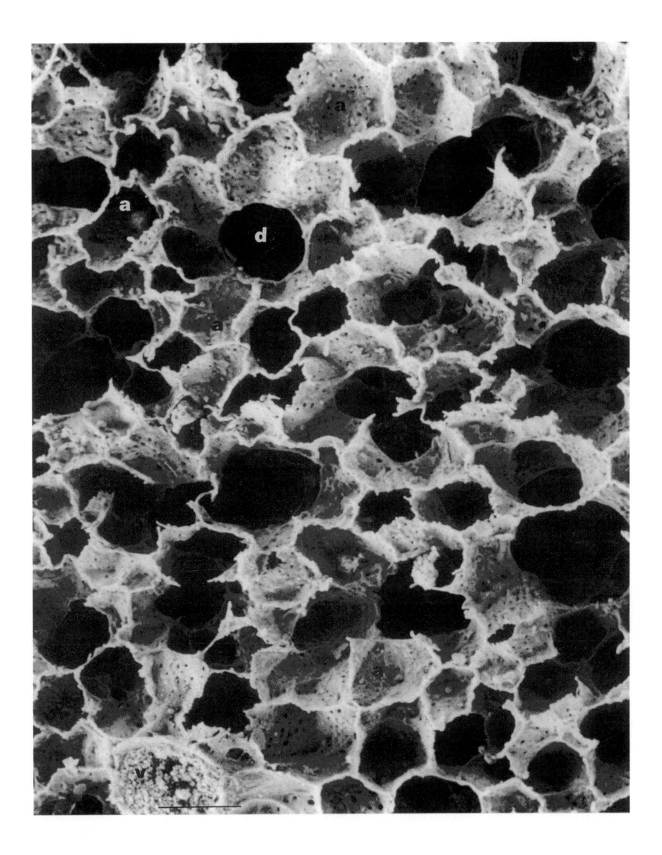

Fig. 86 Scanning electron micrograph of parenchyma of lung of the vervet monkey *Cercopithecus aethiops* showing respiratory duct (d) and alveoli (a). v, blood vessel. Scale bar, 0.13 mm.

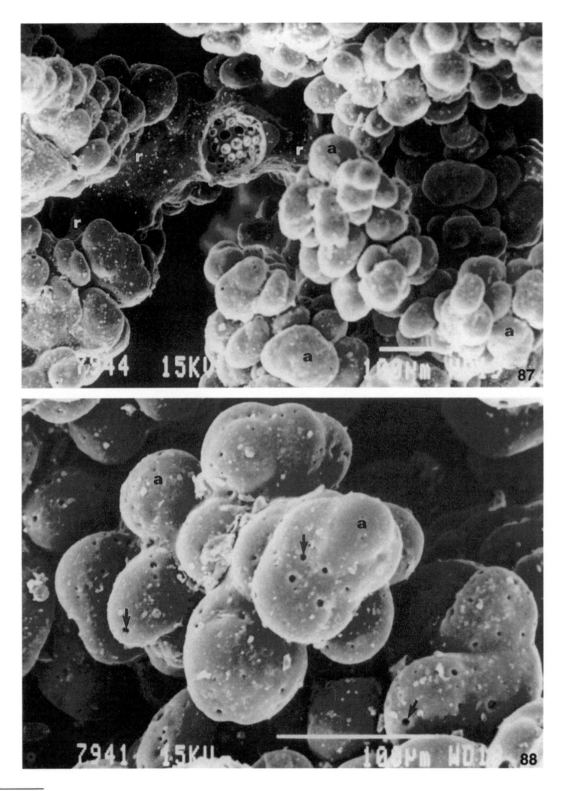

Fig. 87 Scanning electron micrograph of a preparation made by injection of latex rubber into the airways of lung of the domestic pig *Sus scrofa* showing respiratory bronchioles (r) giving rise to a grapelike cluster of alveoli (a). Scale bar, 100 μm.

Fig. 88 Scanning electron micrograph of a preparation made by injection of latex rubber into the airways of lung of the domestic pig *Sus scrofa* showing alveoli (a). Arrows, interalveolar pores. Scale bar, 100 μm.

Fig. 89 Scanning electron micrograph showing bronchial (luminal) surface of lung of the vervet monkey *Cercopithecus aethiops* showing ciliated epithelial lining cells (c) and secretory cells (s). Scale bar, 10 μm.

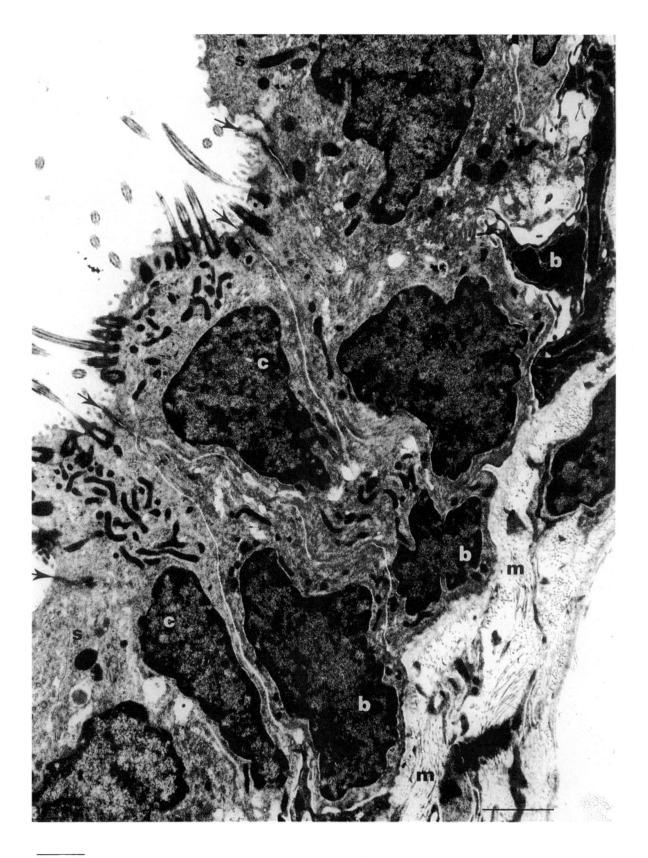

Fig. 90 Transmission electron micrograph of bronchial (luminal) epithelium of lung of the vervet monkey *Cercopithecus aethiops* showing ciliated epithelial cells (c) and secretory cells (s). b, basal cells; m, basement membrane; arrows, intercellular junctions. Scale bar, 2.6 μm.

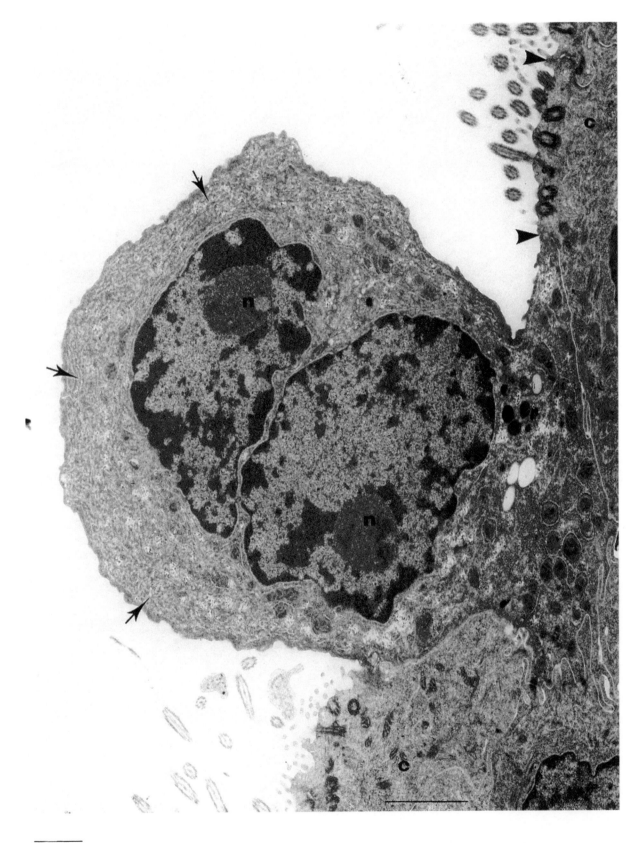

Fig. 91 Transmission electron micrograph of respiratory epithelium of lung of a bat *Phyllostomus hastatus* showing Clara cell (arrows). Arrowheads, intercellular junctions between ciliated cells (c) and the Clara cell; n, nucleus. The actual function of the Clara cells is unknown. Scale bar, 1 μm.

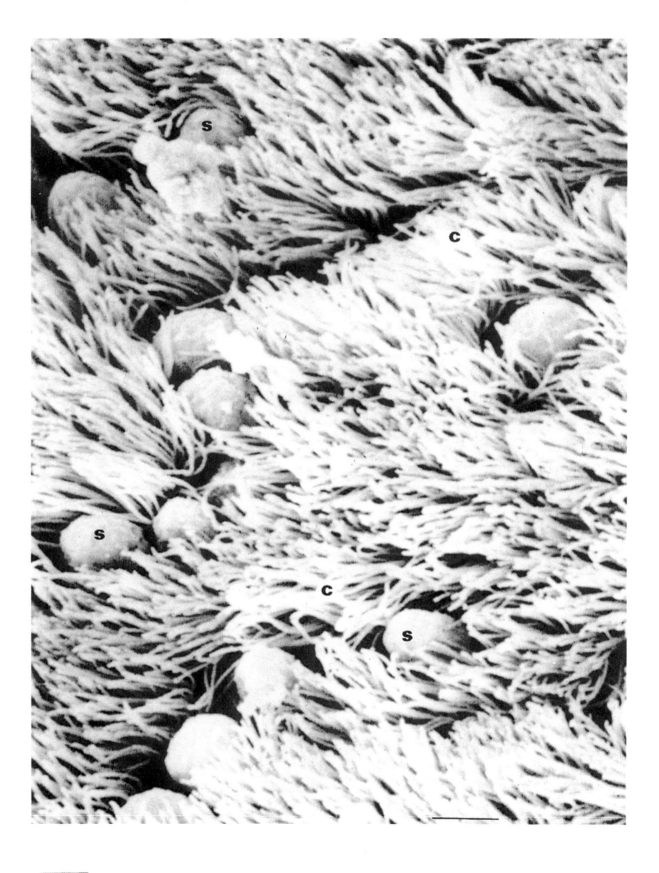

Fig. 92   Scanning electron micrograph showing epithelial lining of trachea of the baboon *Papio anubis* showing ciliated cells (c) and columnar secretory cells (s). Scale bar, 15 μm.

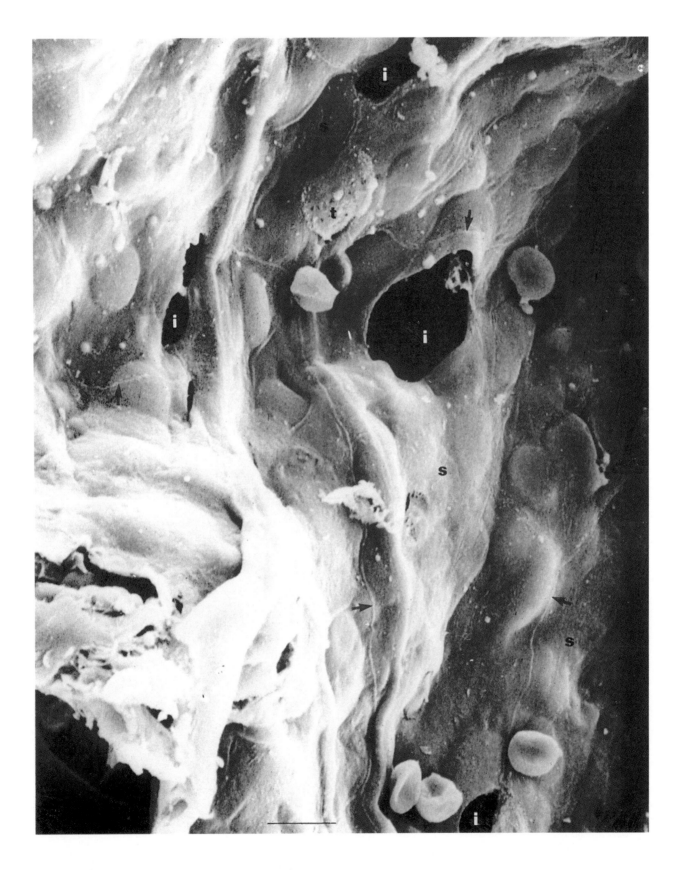

Fig. 93 Scanning electron micrograph showing alveolar (luminal) surface of lung of the baboon *Papio anubis* showing Type I cells (s) and Type II cells (t). Arrows, intercellular junctions between Type I and II cells and between Type I cells. i, interalveolar pores. Scale bar, 10 μm.

Fig. 94 Scanning electron micrograph showing alveolar (luminal) surface of lung of the baboon *Papio anubis* showing blood capillaries (c) bulging into air space. Arrowheads, intercellular junctions of Type I cells. Scale bar, 0.4 µm.

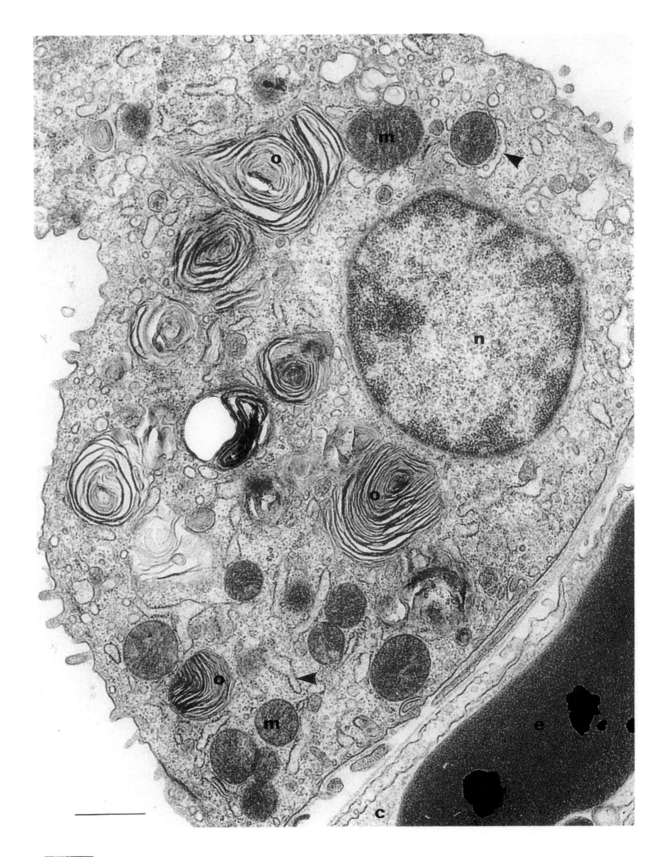

Fig. 95 Transmission electron micrograph showing Type II cell of lung of the bat *Miniopterus minor* showing lamellated osmiophilic bodies (o) and mitochondria (m). Arrowheads, rough endoplasmic reticulum; n, nucleus; e, erythrocyte contained in a blood capillary (c). Scale bar, 2.7 μm.

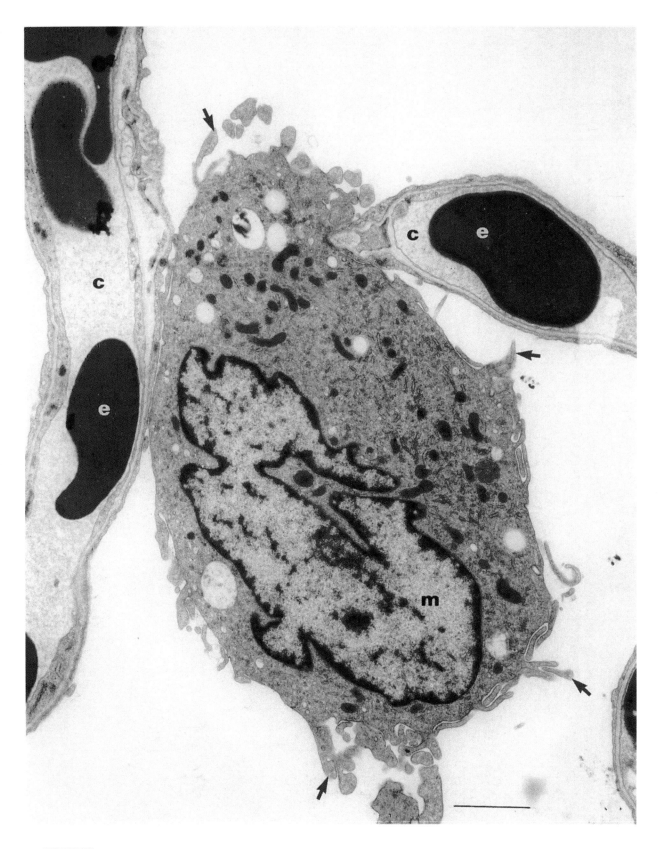

Fig. 96   Transmission electron micrograph showing surficial macrophage (m) of lung of the bat *Miniopterus minor* passing through an interalveolar pore. Arrows, filopodia; c, blood capillaries; e, erythrocytes. Scale bar, 5 μm.

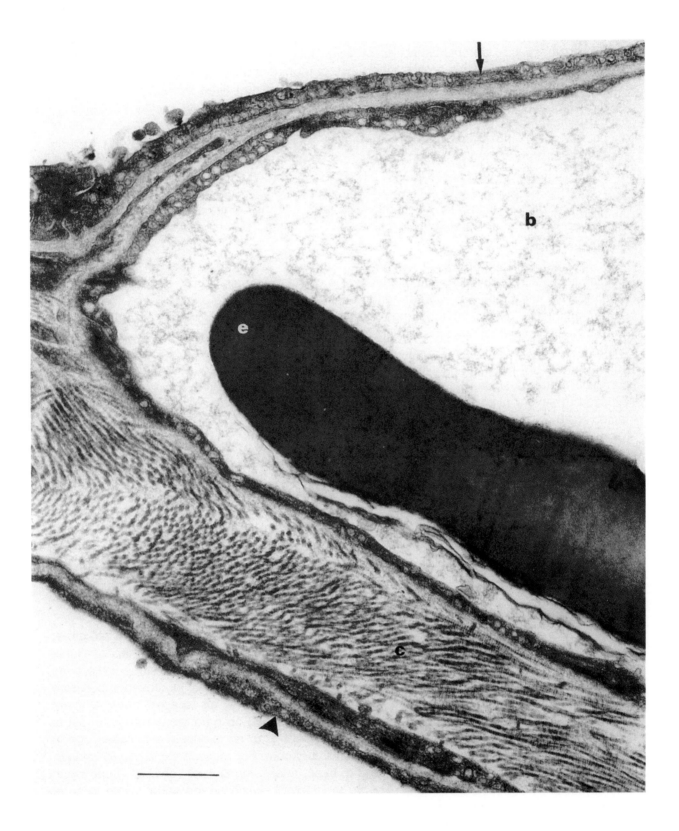

Fig. 97 Transmission electron micrograph showing interalveolar septum of lung of the lesser bushbaby *Galago senegalensis* showing respiratory aspect (arrow) and supporting aspect (arrowhead). The interstitial space of the thick part contains collagen fibers (c). e, erythrocyte contained in a blood capillary (b). Scale bar, 0.5 μm.

of the Eocene that are 50 million years old. Morphologically, they resemble the modern Microchiroptera. The protobats may have been tree-dwelling omnivores that began gliding between trees while foraging.

The mass-specific aerobic capacities of flying bats are equal to those of forward flapping birds. They are 2.5 to 3 times greater than those of running mammals of the same body mass. Bats can increase their oxygen consumption during sustained flight by factors of 20 to 30 times. Under an ambient temperature of 20°C, a 12 gram bat (*Myotis velifer*) increases its oxygen consumption by an astounding factor of 130. Bats are excellent fliers in terms of speed, distance, and maneuverability. Moreover, at least experimentally, they can tolerate extreme hypoxia. A speed of 64 kph has been reported in *Eptesicus fuscus. Lasiurus borealis, L. cinereus, Lasionycteris noctivagans, Nyctalus noctula,* and *Tadarida brasiliensis* migrate over distances of about 1,000 km. As they forage, *Epomophorus wahlbergi* and *Scotophilus viridis* may cover distances of about 500 km in one night. Regarding agility and maneuverability, small bats show great scope for flight, with some species capable of hovering.

Compared with the lung-air sac system of birds (within the existing structural and functional limitations intrinsic to the overall design of the mammalian lung, e.g. tidal ventilation, relatively thicker blood-gas barrier, and less extensive respiratory surface), bats have morphometrically highly refined lungs. Together, adaptive physiological parameters and morphometric specializations have greatly increased the uptake and transport of oxygen by the chiropteran cardiopulmonary system. A characteristic mammalian lung has been exquisitely honed to attain a diffusing capacity of oxygen during flight at rates equal to those of the more complex, "better designed" bird lungs [Chapter 8]. For example, during flight or at rest, when exposed to severe hypoxic stress, *Phyllostomus hastatus* can maintain an oxygen extraction factor of 20%, a value comparable to that of a bird in flight.

The main pulmonary morphological and physiological parameters in the chiropteran respiratory adaptive portfolio are large heart, large lung, high venous hematocrit, and high hemoglobin concentration. With an extensive respiratory surface, large pulmonary capillary blood volume, and thin blood-gas barrier, the bat lung has a higher morphometric diffusing capacity for oxygen compared to that of a nonvolant mammal. The lungs of bats practically fill the coelomic cavity. Atypical of mammals, the gastrointestinal system is both small and poorly morphologically differentiated into different functional regions. The transit time of food in the gastrointestinal system is short. The mass-specific respiratory surface area of 138 $cm^2 g^{-1}$ in the epauletted fruit bat *Epomophorus wahlbergi,* is the highest value reported thus far in a vertebrate lung. The thinnest blood-gas barrier in a mammalian lung (0.120 μm) has been reported in *Phyllostomus hastatus.*

From the perspective of engineering, to increase the internal surface area of a structure, overall enlargement of the containing space and/or increased subdivision of the available space can be effected. For definite reasons, bats have adopted the first strategy while birds [Chapter 8] have applied the second. In the compliant mammalian lung, increased subdivision of the parenchyma generates small alveoli. Narrow terminal air spaces have a great propensity for collapsing from high surface tension forces and obligate greater energy cost to inflate. Conceivably, bats "opted" for a large lung rather than an intensely subdivided one in order to optimize gas exchange by reducing ventilatory cost. For reasons explained in Chapter 8, regarding the compact parabronchial lung of birds, after relegation of ventilatory function to the air sacs, it was feasible and cost-effective to intensely subdivide the parenchyma in order to increase the repiratory surface area. A one-to-one accord between wing beat and breathing cycles occurs in all the bats studied to date. In birds, though flight muscles attach directly onto the sternum, such a synchrony occurs only in a few species. The harmony between wing beat and respiratory cycles must allow bats to ventilate their lungs at a lower energy cost and probably more efficiently. Developing an intensely divided lung (i.e., one with small alveoli) would have negated all the advantages derived by the synchronization between breathing and wing movements.

10

Summary and Conclusions

Energy is vital for life. In economic terms, it is the currency with which living things "purchase necessary goods and services". Production, storage, and optimal use of energy are important processes for supporting the activities that sustain life. Many energy transactions that occur in living systems are chemical in nature. The sun is the primary source of energy. The leaves of green plants are the receptors and transducers of sunlight. Solar energy is captured by plants and conserved in chemical bonds of compounds such as glucose, fatty acids, starch, and glycogen. In the physical and natural worlds, from atomic to ecological levels, the acquisition and utilization of energy follows Maxwell's laws of thermodynamics. Regarding the first law, the overall measure of energy in the Universe is constant: no more can be created nor can the existing amount be destroyed. It can, however, be converted from one form to another. According to the second law, the freely accessible energy of a system (the energy available to do work) decreases inexorably while entropy increases. Disorder (= chaos) is hence the unconstrained state of matter. Inasmuch as living systems are complex, highly organized, stable states of matter that not only maintain their integrity (though temporarily), but also refine themselves, it would appear on a casual glance that life "runs uphill in a downhill Universe", i.e., it advances counter to the dissipation of energy. If that actually occurred, the second law of thermodynamics would be violated. Instead of jeopardizing life, however, the second law of thermodynamics de facto guarantees it. Unlike the closed thermostatic state of ordinary laboratory (controlled) chemical reactions, living systems are open thermodynamic states of matter that access energy from their immediate surroundings as well as the Universe at large. As long as energy is available to be shifted from its proximate environment into a living entity, the biotic part of the system can increase in organizational complexity while still advancing the increasing entropy of the system.

Animals are conduits of energy in the immensity of the Universe. Competition for diminishing resources (specifically the amount of freely available energy to do work) is thought to be the propelling factor in the processes of evolution by natural selection and adaptation. Intricate interdependence in the harnessing of raw materials, use, and conservation of energy occurs between animals and plants and between animals themselves. Operating under finite reserves, it is imperative for animals to optimize the structure and performance of their constitutive body systems. Through the evolutionary continuum, morphological designs, particularly those of the respiratory organs, have been honed, eliminating or reducing extravagant expenditure of energy on redundant or superfluous structures. This has averted extravagant support of underutilized or unused capacities, providing optimal functional and structural conditions and states.

The designs of the respiratory and the cardiovascular systems (highly integrated organic assemblages that comprehensively acquire and deliver oxygen to the tissue cells for production of energy by oxidative phospholylation) are fundamental to the wholesomeness of vertebrate life. This is clearly demonstrated by the fact that during embryological development, the two systems are the very first to form and become functionally competent. Respiratory organs have evolved tractably with needs for molecular oxygen to support determinate metabolic capacities. Quantitatively, the structure of a respiratory organ is highly malleable to functional demands. Based on reasonably similar design prescripts, a wide spectrum of respiratory structures have evolved in vertebrates. Solutions to common respiratory demands are typified by profound convergence of design, structure, and refinement. Compared with large, indolent, ectothermic-heterotherms, small endothermic homeotherms with particularly high aerobic capacities have very well-refined respiratory organs. Compared to phylogenetic levels of development, metabolic demands for molecular oxygen have set to a greater extent, the designs and degrees of structural refinement of the gas exchangers. The dictum that "in biology there are no rules but only necessities" is epitomized in the design and fabrication of the respiratory organs. For instance, from the perspectives of the construction and efficiency of the human lung, the design is far from having attained the pinnacle of evolved vertebrate respiratory organs. A practical point that clearly demonstrates this is the fact that while only healthy, well-acclimated humans can survive at the top of Mt. Everest without a supplemental source of oxygen, birds fly freely over the mountain at much higher altitudes. For example, during its annual migration the bar-headed goose (*Anser indicus*) flies directly (without acclimation) from the Indian subcontinent (an altitude of about sea level) to Central Asia to escape winter cold and to breed. It has been estimated that if human beings possessed the avian parabronchial lung instead of a bronchoalveolar lung, they would be able to ascend to much higher elevations without need for an auxilliary oxygen source.

The fundamental structural features of a respiratory organ include: an extensive respiratory surface area, a thin water/air-blood (tissue) barrier, and a large volume of capillary blood. In gills, an extensive surface area is created by a stratified structural design, while in lungs it is achieved by internal subdivision. Partitioning of the lung yields small terminal respiratory units highly susceptible to collapse from surface tension forces at the air-water interface and entail greater effort to ventilate. The surfactant, a phospholipid lining, lowers surface tension forces and decreases the energetic cost of ventilating the lung. In birds, by dissociating the lung (the gas exchanger) from the ventilatory apparatus (air sacs) and developing a compact, firmly fixed, nonexpansile lung, it was possible for the parenchyma (the periparabronchial gas-exchange tissue) to be intensely subdivided. That yielded remarkably narrow air capillaries while enhancing the respiratory surface area in a small lung. In the compliant bat lung, the respiratory surface area was increased by striking enlargement of the lung rather than by increased compartmentalization of the parenchyma, a process that would increase the energetic cost of ventilating the lung.

To a great extent, a respiratory organ or structure is defined by the manifest design rather than by the properties and composition of the components that constitute it. A simple, cell membrane is the most elementary but practical respiratory structure. After billions of years, the Protozoa still utilize it! In its most highly reduced form, a model of a gas exchanger comprises a fabrication wherein two fluid media are in close proximity and a PO_2 exists between them: the site of the body where this occurs and the tissues that constitute it are irrelevant. To maintain a partial pressure gradient, i.e., a driving force; the respiratory structure must be ventilated with water/air and perfused with blood. The nonprescriptive nature of the structure of the respiratory organs explains the remarkable diversity of the evolved organs. In vertebrates, these include swim bladders, gills, skin, gastrointestinal system, buccal cavity, and lungs. Though performing a similar function, embryologically these structures develop from different germ cells. Utilizing the different infrastructural resources with which they are individually phylogenetically endowed, different animals have evolved various respiratory structures and in some cases, through remarkable convergence, developed congruous ones. For example, in bats and birds, the only two extant volant vertebrates (animals separated by some 100 million years of evolution), using distinct strategies, different but equally efficient respiratory organs have developed. In bats, the "inferior" mammalian lung has been greatly morphometrically honed and the respiratory efficiency augmented by a suite of physiological (cardiovascular) adaptations such as high venous hematocrit and hemoglobin concentration. Respiratory organs of equivalent respiratory efficiency have developed in birds and bats to support a common lifestyle—flight, a very energy-costly mode of locomotion.

Gills (evaginated gas exchangers) are the primordial respiratory organs that developed for water breathing and aquatic existence while lungs (invaginated gas exchangers) developed for air breathing. Surviving at the water-air interface, transitional (= bimodal = amphibious) breathers use specialized organs to extract oxygen from water and air. The design and construction of the modern gas exchangers have occurred through trial-and-error and cost-benefit analysis so as to ensure optimal design and performance. This rigorous "quality control" process has decreed that only a limited number of structurally feasible and functionally competent forms and designs eventuate. Trade-offs and compromises have been transacted in the development of the ultimate gas exchangers. For example, while invagination of the gas exchangers averted risk of desiccation through excessive water loss, such respiratory structures could only be tidally (= bidirectionally) ventilated, a less efficient mechanism for maintaining high PO_2 at the respiratory surface. On the other hand, gills (evaginated gas exchangers) could be continuously and unidirectionally ventilated, allowing the highly efficient countercurrent disposition between water and blood. Such a highly efficient system was necessary for survival in water, a medium that, compared to air, is deficient in oxygen. The analogy in the design and construction of the gas exchangers bespeaks the fact that, to a great extent, corresponding selective pressures prescribed the ultimate forms. Moreover, there are only two naturally occurring respirable fluid media-water and air. Gas exchangers have evolved to utilize one or the other and in rarer events (in the case of bimodal breathers) both.

Factors that have profoundly shaped the structure and function of vertebrate respiratory organs are: body mass, respiratory medium utilized, phylogenetic level of development, lifestyle, and habitat. The different physicochemical properties of water and air have particularly impacted on the structural differences between gills and lungs. The differences are so great that gills will not effectively function in air and neither will lungs in water. The presentation and exposure of the respiratory media (water/air to blood), features determined by the geometry and arrangement of the structural components, contribute highly to respiratory efficiency. In this respect, the countercurrent presentation between water and blood in the fish gills is the most efficient design to have evolved in the gas exchangers. The design was imperative for survival in water—a medium that contains relatively less oxygen, one in which the diffusivity of oxygen is slower, and given to its greater viscosity is more expensive to breathe. In the respiratory organs of air-breathing vertebrates, the exposure of capillary blood to air is best perfected in the diffuse arrangement of the air capillaries of the avian lung. The air and blood capillaries intertwine, providing optimum exposure of blood to air: the blood is literally suspended in air. A double capillary design occurs in the lungs of lungfishes and amphibians and generally in those of reptiles while a single capillary design occurs in the adult mammalian lung. Capillary loading (ratio of volume of capillary blood to respiratory surface area) in lungs with a double capillary arrangement is high. Since a large volume of blood is exposed to air across a limited surface area, this manifests a poor design. The low capillary loading that characterizes the single capillary system of the mammalian lung indicates much better exposure of blood to air: an extensive surface is available for gas exchange. Fractal geometry is an important feature in the design and construction of the respiratory organs. The highly versatile design scheme allows the respiratory organs to scale and function optimally under different states, conditions, and circumstances without failure. Moreover, congruous morphologies are maintained over a wide range of body size, shape, and metabolic capacities.

Unlike human-made machines that are normally built to perform one task best, biological structures (including respiratory organs) are dynamic, multifunctional, composite entities. Gills perform important nonrespiratory functions such as ionic exchange, hormonal metabolism, and ammonia excretion while lungs participate in pH homeostasis and production and metabolism of vital pharmacologically active chemical factors. The designs and constructions of respiratory organs must be seen and understood in the context of the multiplicity of their roles and functions and not simply from the narrow one of gas exchange.

References and Works to Consult

1 FUNDAMENTAL PRINCIPLES AND CONCEPTS

Adelman, R., R.L. Saul, and B.N. Ames, 1988. Oxidative damage to DNA: relation to species metabolic rate and lifespan. Proc. Natl. Acad. Sci. USA 85: 2706-2708.

Alberch, P. 1980. Ontogenesis and morphological diversification. *Amer. Zool.* 20: 653-667.

Alexander, R. McN. 1985. The ideal and the feasible: physical constraints on evolution. *Biol. J. Linn. Soc.* 26: 345-358

Alexander, R. McN. 1996. Optima for Animals. Princeton University Press, Princeton, NJ.

Änggård, E .1975. Biosynthesis and metabolism of prostaglandins in the lung, pp. 301-311. In: A.E. Junod and R. Haller [eds.]. Lung Metabolism. Academic Press, NY, USA.

Atchley, W.R. and B.K. Hall. 1991. A model for development and evolution of complex morphological structures. *Biol. Rev.* 66: 101-157.

Bakhle, Y.S. 1975. Pharmocokinetic function of the lung, pp. 293-299. In: A.F. Junod and R. Haller [eds.]. Lung Metabolism. Academic Press, NY, USA.

Balter, M. 1996. Looking for clues to the mystery of life on Earth. *Science* 273: 870-872.

Barel, C.D.N. 1993. Concepts of an architectonic approach to transformation morphology. *Acta Biotheoretica* 41: 345-381.

Barel, C.D.N., G.C. Anker, F. Witte, R.S.C. Hoogerhoud, and T. Goldschmidt. 1989. Constructional constraint and its ecomorphological implications. *Acta Morphol. Neerl. Scand.* 27: 83-109.

Barman, S.A., L.L. McCloud, S.D. Catravas and I.C. Ehrhart. 1996. Measurement of pulmonary blood flow by fractal analysis of flow heterogeneity in isolated canine lungs. *J. Appl. Physiol.* 81: 2039-2045.

Barnsley, M.F., P. Massopust, H. Strickland and A.D. Sloan. 1987. Fractal modeling of biological structure and function. *Ann. NY Acad. Sci.* 504: 179-194.

Bar-Nun, A., and A. Shaviv. 1975. Dynamics of the chemical evolution of the Earth's primitive atmosphere. *Icarus* 24: 197-210.

Barthelemy, L. 1987. Oxygen poisoning, pp. 152-162. In: P. Dejours [ed.]. Comparative Physiology of Environmental Physiology, vol 2. Karger, Basel.

Bartholomew, G.A. 1982. Scientific innovation and creativity: a zoologists point of view. *Amer. Zool.* 22: 227-235.

Bartholomew, G.A. 1982. Energy metabolism. pp. 57-110. In: M.S. Gordon MS [ed.]. Animal Physiology: Principles and Adaptations. MacMillan, NY USA.

Basalla, G. 1989. The Evolution of Technology. Cambridge University Press, Cambridge.

Bassingthwaighte, J.B., L.S. Liebovitch, and B.J. West. 1994. Fractal Physiology. Oxford University Press, Oxford.

Baum, D.A., and A. Larson. 1991. Adaptation reviewed: A phylogenetic methodology for studying character macroevolution. *Syst. Zool.* 40: 1-18.

Beament, J.W.L. 1960. Physical models in biology, pp. 66-123. In: J.W.L. Beament [ed.]. Models and Analogues in Biology. Symp. Soc. Exper. Biol. No. 14. Academic Press, NY USA.

Bennett, A.F. and W.R. 1988. Structural and functional determinates of metabolic rate. *Amer. Zool.* 28: 699-708.

Bennett, A.F. and W.R. Dawson. 1976. Metabolism, pp. 127-223. In: C. Gans and W.R. Dawson [eds.]. Biology of Reptilia, vol 5. Academic Press, NY USA.

Bennett, A.F. and J.A Ruben. 1979. Endothermy and activity in vertebrates. *Science* 206: 649-655.

Berker, L.V. and L.C. Marshall. 1965. The origin and rise of oxygen concentration in the Earth's atmosphere. *J. Atmos. Sci.* 22: 225-261.

Blum, H.F. 1955. Time's Arrow and Evolution. Princeton University Press, Princeton, NJ USA.

Bock, W.J. and G. von Wahlert. 1965. Adaptation and the form-function complex. *Evolution.* 19:269-299.

Bonner, J.T. 1988. The Evolution of Complexity. Princeton University Press, NJ USA.

Boulière, F. 1975. Mammals, small and large: the ecological implication of size, pp. 1-8. In: F.B. Golley and K. Petrusewicz [eds.]. Small Mammals: Their Productivity and Population Dynamics. Cambridge University Press, Cambridge.

Bray, A.A. 1985. The evolution of the terrestrial vertebrates: environmental and physiological considerations. *Phil. Trans. R. Soc. Lond.* 309B: 289-322.

Briggs, J. 1992. Fractals: The Pattern of Chaos. Simon and Schuster, NY USA.

Broda, E. 1978. The Evolution of Bioenergetic Processes. Pergamon, Oxford.

Brown, A.C. 1993. Variability in biological systems. *S. Afr. J. Sci.* 89: 308-309.

Brown, A.C. 1994. Is biology science? *Trans. R. Soc. SA.* 49: 141-146.

Brown, J.H., P.A. Marquet, and M.L. Taper. 1993. Evolution of body size: consequences of an energetic definition of fitness. *Amer. Nat.* 142: 573-584.

Burggren, W.W. 1991. Does comparative respiratory physiology have a role in evolutionary biology (and vice versa)? pp. 1-13. In: A.J. Woakes, M.K. Grieshaber, and C.R. Bridges [eds.]. Physiological Strategies for Gas Exchange and Metabolism. Cambridge University Press, Cambridge.

Burggren, W.W. and A.W. Pinder. 1991. Ontogeny of cardiovascular and respiratory physiology in lower vertebrates. *Ann. Rev. Physiol.* 53: 107-135.

Burggren, W.W. and J. Roberts. 1991. Respiration and metabolism, pp. 353-435. In: C.L. Prosser [ed.]. Environmental and Metabolic Animal Physiology. Wiley, Liss, NY USA.

Calder, W.A. 1984. Size, Function and Life History. Harvard University Press, Cambridge, MA USA.

Calvin, M. 1956. Chemical evolution and the origin of life. *Amer. Sci.* 44: 248-263.

Cameron, J.C. 1989. The Respiratory Physiology of Animals. Oxford University Press, Oxford.

Carroll, R.L. 1988. Vertebrate Palaeontology and Evolution. WH Freeman and Co., NY USA.

Canfield, D.E. and A. Teske. 1996. Late Proterozoic rise in atmospheric oxygen concentration inferred from phylogenetic and sulphur-isotope studies. *Nature, Lond.* 382: 127-132.

Carter, G.S. 1967. Structure and Habitat in Vertebrate Evolution. Washington University Press, Seattle, WA, USA.

Chapman, D.J. and M.A. Ragan. 1980. Evolution of biochemical pathways: evidence from comparative biochemistry. *Ann. Rev. Pl. Physiol.* 31: 639-678.

Chapman, D.J. and J.W. Schopf. 1983. Biological and biochemical effects of the development of an aerobic environment, pp. 302-320. In: J. W. Schopf [ed.]. Earth's Earliest Atmosphere: Its Origin and Evolution. Princeton University Press, Princeton, NJ USA.

Charnov, E.I. 1997. Trade-off-invariant rules for evolutionary stable state life histories. *Nature, Lond* 387: 393-394.

Chen, C.,B. Rabourdin, and C.S. Hammen. 1987. The effect of hydrogen sulfide on the metabolism of *Solemya velum* and enzymes of sulfide oxidation in gill tissue. *Comp. Biochem. Physiol.* 88B: 949-952.

Childress, J.J., and H. Felbeck, and G.N. Somero. 1989. Symbiosis in the deep sea. *Sci. Amer.* 256: 107-112.

Clark, J.M., and C.J. Lambertsen. 1971. Pulmonary oxygen toxicity: a review. *Pharmacol. Rev.* 23: 37-98.

Cloud, P. 1983. Early biogeologic history: the emergence of a paradigm, pp. 14-31. In: J.W. Schopf [ed.]. Earth's Earliest Biosphere: Its Origin and Evolution. Princeton University Press, Princeton, NJ USA.

Cochrane, C.G. 1991. Cellular injury by oxidants. *Am. J. Med.* 91: 23S-30S.

Cohen, J. 1995. Getting all turned around over the origins of life on Earth. *Science* 267: 1265-1266.

Colbert, E.H., and M. Morales. 1991. Evolution of the Vertebrates, 4th ed. Wiley-Liss, NY USA.

Conway-Morris, S. 1993. The fossil record and the early evolution of Metazoa. *Nature, Lond* 361: 219-225.

Coope, A., and D. Penny. 1997. Mass survival across the Cretaceous-Tertiary boundary: Molecular evidence. *Science* 275: 1109-1113.

Crapo, J.D. 1987. Hyperoxia: Lung injury and localization of antioxidant defenses, pp. 163-176. In: P. Dejours [ed.]. Comparative Physiology of Environmental Adaptations, vol 2. Karger, Basel.

Crompton, A.W., C.R. Taylor, and J.A. Jagger. 1978. Evolution in homeothermy in mammals. *Nature, Lond* 272: 333-336.

Csillag, C., and P. Aldhous. 1992. Signs of damage by radicals. *Science* 258: 1875-1876.

Curtis, S.E., J.T. Peek, and D.R. Kelly. 1993. Partial liquid breathing with perflurocarbon improves arterial oxygenation in acute canine lung injury. *J. Appl. Physiol.* 75: 2696-2702.

Dejours, P. 1981. Principles of Comparative Physiology, 2nd ed. Elsevier North-Holland, Amsterdam.

Dejours, P. 1988. Respiration in Water and Air: Adaptations, Regulation and Evolution. Elsevier North-Holland, Amsterdam.

Denney, M.W. 1993. Air and Water: The Biology and Physics of Life's Media. Princeton University Press, Princeton, NJ, USA.

Dobzhansky, T. 1973. Nothing in biology makes sense except in the right of evolution. *Amer. Biol. Teacher.* 35: 125-129.

Dullemeijer, P. 1974. Concepts and Approaches in Animal Morphology. Van Corkum, Assen.

Duncker, H.R. 1978. General morphological principles of amniotic lungs, pp. 1-15. In: J. Piiper [ed.]. Respiratory Function in Birds, Adult and Embryonic. Springer, Berlin.

Duncker, H.R. 1991. The evolutionary biology of homoiothermic vertebrates: the analysis of complexity as a specific task of morphology. Verh. *Dtsch. Zool. Ges.* 84: 39-60.

Dyer, B.D., and R.A. Ober. 1994. Tracing the History of the Eukaryotic Cells. Columbia University Press, NY, USA.

Economos, A.C. 1979. Gravity, metabolic rate, and body size of mammals. *Physiologist* 22: S71.

Else, P.L., and A.J. Hubert. 1981. Comparison of the 'mammal machine' and the 'reptilian machine' energy production. *Amer. J. Physiol.* 240: R3-R9.

Erwin, D.H. 1996. The mother of mass extinctions. *Sci. Amer.* 275: 56-62.

Farhi, L., and H. Rahn. 1955. Gas stores of the body and the steady state. *J. Appl. Physiol.* 7: 472-484.

Felbeck, H., J.J. Childress, and G.N. Somero. 1981. Calvin-Benson cycle and sulphide oxidation enzymes in animals from sulphide-rich habitats. *Nature, Lond* 293: 291-293.

Fenchel, T., and B.J. Finlay. 1994. The evolution of life without oxygen. *Amer. Natur.* 82: 22-29.

Fishman, A.P. 1983. Comparative biology of the lung. *Amer. Rev. Respir. Dis.* 128: S90-S91.

Fontana, W., and L.W. Buss. 1993. The arrival of the fitness: Toward a theory of biological organization.Technical Report No. SFI 93-09-055, pp. 1-71. Santa Fe Institute, Santa Fe.

Forey, P., and P. Janvier. 1994. Evolution of the early vertebrates. *Amer. Sci.* 82: 554-565.

Forman, H.J., and A.B. Fisher. 1981. Antioxidant defense, pp. 235-249. In: D. L. Gilber [ed.]. Oxygen and Living Processes: An Interdisplinary Approach. Springer-Verlag, NY USA.

Frakes, L.A. 1979. Climates throughout Geologic Time. Elsevier Publishing, Amsterdam.

Fung, Y.B. 1993. Biomechanics: Mechanical Properties of Living Tissues, 2nd ed. Springer Verlag, Berlin-Heidelberg.

Fustec, A., D. Desbruyères, and S.K. Juniper. 1987. Deep-sea hydrothermal vent communities at 13°N on the East Pacific rise: microdistributions and temporal variations. *Biol. Oceanography.* 4: 121-164.

Futuyuma, D.J. 1986. Evolutionary Biology, 2nd ed. Sinauer Associates Inc., Sunderland, MA USA.

Gans, C. 1988. Adaptation and the form-function relation. *Amer. Zool.* 28: 681-697.

Gardiner, B.G. 1980. Tetrapod ancestry: a reappraisal, pp. 135-207. In: A.L. Panchen [ed.]. The Terrestrial Environment and the Origin of Land Vertebrates. Academic Press, London.

Garland, T., and R.B. Huey. 1987. Testing symmorphosis: Does structure match functional requirements? *Evolution* 41: 1404-1409.

Gillis, A.M. 1991. Can organisms direct their evolution? *BioScience* 41: 202-205.

Glazier, J.B., J.M.B. Hughes, J.E. Maloney, and J. B. West. 1967. Vertical gradient of alveolar size in lungs of dogs frozen intact. *J. Appl. Physiol.* 23: 694-705.

Goldberger, A.L., and B.J. West. 1987. Fractals in Physiology and Medicine. *Yale. J. Biol. Med.* 60: 421-435.

Gordon, J.E. 1978. Structure; or, Why Things Dont Fall Down. Plenum Press, NY USA.

Gould, S.J. 1994. The evolution of life on Earth. *Sci. Amer.* 271: 63-69.

Grassle, J.F. 1985. Hydrothermal vent animals: distribution and biology. *Science* 229: 713-717.

Gutmann, W.F. 1977. Phylogenetic reconstruction: theory, methodology, and application to chordate evolution, pp. 45-96. In: M.K. Hecht, P.C. Goody, and B.M. Hecht [eds.]. Major Patterns of Vertebrate Evolution. NATO Adv Study Inst Series, vol 14. Plenum Press, NY USA.

Hayes, B. 1994. Nature's algorithms. *Amer. Sci.* 82: 206-210.

Hildebrand, M., D.M. Bramble, K.F. Liem, and D.B. Wake [eds.]. 1985. Functional Vertebrate Morphology. Harvard University Press, Cambridge MA USA.

Hinkle, P.C., and R.E. McCarty. 1978. How cells make ATP. *Sci. Amer.* 238: 104-123.

Hochachka, P.W. 1980. Living without Oxygen: Closed and Open Systems in Hypoxia Tolerance. Havard University Press, Cambridge, MA USA.

Hochachka, P.W. 1979. Cell metabolism, air breathing, and the origins of endothermy, pp. 253-288. In: S.C. Wood, C. Lenfant [eds.]. Evolution of Respiratory Processes: A Comparative Approach. Marcel Dekker Inc., NY USA.

Hochachka, P.W., J. Fields, and T. Mustafa. 1973. Animal life without oxygen: basic biochemical mechanisms. *Amer. Zool.* 13: 543-555.

Hochachka, P.W., and G.N. Somero. 1973. Strategies of Biochemical Adaptations. Saunders, Philadelphia, PA USA.

Hughes, G.M. 1965. Comparative Physiology of Vertebrate Respiration, 2nd ed. Heineman, London.

Iwai, T., and I. Nakamura. 1964. Branchial skeleton of the bluefin tuna, with special reference to the gill rays. *Bull. Misaki. Mar. Biol. Inst., Kyoto. Univ.* 6: 21-25.

Jahnke, L., and H.P. Klein. 1979. Oxygen as a factor in eukaryocyte evolution: some effects of low levels of oxygen on Saccharomyces cerevisiae. *Origins of Life* 9: 329-334.

Janis, C. 1993. Victory by default: the mammalian succession, pp. 169-217. In: S.J. Gould [ed.]. The Book of Life. Ebury-Hutchison, London.

Jarvik, E. 1980. Basic Structure and Evolution of Vertebrates, vol 2. Academic Press, London.

Johansen, K. 1971. Comparative physiology: gas exchange and circulation in fishes. *Annu. Rev. Physiol.* 33: 569-612.

Johansen, K. 1972. Heart and circulation in gill, skin and lung breathing. *Respir. Physiol.* 14: 193-210.

Jones, J.D. 1972. Comparative Physiology of Respiration. Edward Arnold, London.

Kardong, K.V. 1995. Vertebrate: Comparative Anatomy, Function and Evolution. Wm. C. Brown Publishers, Dubuque, IA USA.

Kasting, J.F., S.C. Liu, and T.M. Donahue. 1979. Oxygen levels in the prebiological atmosphere. *J. Geophys. Res.* 84: 3097-3107.

Kleiber, M. 1961. The Fire of Life: An Introduction to Animal Energetics. Wiley, NY USA.

Kleiber, M. 1965. Respiratory exchange and metabolic rate, pp. 927-938. In: W.O. Fenn, and H. Rahn [eds.]. Handbook of Physiology, sect 3, Respiration, vol II. American Physiological Society, Washington DC, USA.

Krogh, A. 1941. The Comparative Physiology of Respiratory Mechanisms. University of Pennsylvania Press, Philadelphia, PA USA.

Kylstra, J.A., C.V. Paganelli, and E. H. Lanphier. 1966. Pulmonary gas exchange in dogs ventilated with hyperbarically oxygenated liquid. *J. Appl. Physiol.* 21: 177-184.

Lauder, G.V. 1981. Form and function: structural analysis in evolutionary morphology. *Paleobiology* 7: 430-442.

Lauder, G.V., and K.F. Liem. 1989. The role of historical factors in the evolution of complex organismal functions, pp. 63-78. In: D. B. Wake, and G. Roth [eds.]. Complex Organismal Functions: Integration and Evolution in Vertebrates. John Wiley and Sons, London.

Lehninger, A.L. 1982. Principles of Biochemistry. Worth, Publishers, NY, USA.

Lefevre, J. 1983. Teleonomical optimization of a fractal model of the pulmonary arterial bed. *J. Theor. Biol.* 102: 225-248.

Liem, K.F., and D.B. Wake. 1985. Morphology: current approaches and concepts, pp. 269-305. In: M. Hildebrand [ed.]. Functional Vertebrate Morphology. Harvard University Press, Cambridge MA USA.

Lindstedt, S.L., and J.H. Jones. 1988. Symmorphosis: The concept of optimal design, pp. 290-310. In: M.E. Feder, A.F. Bennet, W.W. Burggren, and R.B. Huey [eds.]. New Directions in Ecological Physiology. Cambridge University Press, Cambridge.

Maina, J.N. 1994. Comparative pulmonary morphology and morphometry: The functional design of respiratory systems, pp. 111-232. In: R. Gilles [ed.]. Advances in Comparative and Environmental Physiology, vol 20. Springer-Verlag, Heidelberg.

Maina, J.N. 1998. The Gas Exchangers: Structure, Function and Evolution of Respiratory Processes. Springer, Heidelberg.

Maina, J.N. 2000. Is the sheet-flow design a 'frozen core' (a Bauplan) of the gas exchangers? Comparative functional morphology of the respiratory microvascular systems: illustration of the geometry and rationalization of the fractal properties. *Comp. Biochem. Physiol.* 126A :491-515.

Maina, J.N. 2002. The Fundamental Aspects and Features in the Bioengineering of the Gas Exchangers: Comparative Perspectives. Springer, Heidelberg.

Mandelbrot, B.B. 1983. The Fractal Geometry of Nature, 2nd ed. Freeman, San Francisco, CA USA.

Mangum, C.P. 1994. Multiple sites of gas exchange. *Amer. Zool.* 34: 184-193.

Margulis, L. 1970. Origin of Eukaryotic Cells. Yale University Press, New Haven, CT USA.

May, R.M. 1990. How many species? *Phil. Trans. R. Soc. Lond.* 330B: 293-316.

McCutcheon, F.H. 1964. Organ systems in adaptation: the respiratory system, pp. 167-191. In: D. B. Mill, E. F. Adolph, and C.G. Wilber [eds.]. Handbook of Physiology, sect 4, Adaptation to the Environment. American Physiological Society, Washington DC USA.

Muir, B.S., and J.I. Kendall. 1968. Structural modifications in the gills of tunas and some other oceanic fishes. *Copeia* 1968: 388-398.

Nelson, T. R., B. J. West, and A. L. Goldberger 1990. The fractal lung: universal and species related scaling patterns. *Experientia* 46: 251-254.

Osaki, S. 1996. Spider silk as mechanical lifeline. *Nature, Lond.* 384: 419.

Pattle, R. E. 1976. The lung surfactant in the evolutionary tree, pp. 233-255. In: G.M. Hughes [ed.]. Respiration of Amphibious Vertebrates. Academic Press, London.

Panchen, A.L., and T.R. Smithson. 1988. The relationships of the earliest tetrapods, pp. 1-32. In: M.J. Benton [ed.]. The Phylogeny and Classification of the Tetrapods, vol. 1: Amphibians, Reptiles and Birds. Clarendon, Oxford.

Perry, S.F. 1989. Mainstreams in the evolution of vertebrate respiratory structures, pp. 1-67. In: A.S. King, and J. McLelland [eds.]. Form and Function in Birds, vol IV. Academic Press, London.

Perry, S.F. 1992. Evolution of the lung and its diffusing capacity, pp. 142-153. In: J.P.W. Bicudo [ed.]. Vertebrate Gas Transport Cascade Adaptations to Environment and Mode of Life. CRC Press, Boca Raton.

Petroski, H. 1985. To Engineer is Human. St Martin's Press, NY, USA.

Piiper, J., and P. Scheid. 1977. Comparative physiology of respiration: functional analysis of gas exchange organs in vertebrates. *Int. Rev. Physiol.* 14: 219-253.

Piiper, J., and P. Scheid. 1972. Maximum gas transfer efficacy of models for fish gills, avian lungs and mammalian lungs. *Respir. Physiol.* 14: 115-124.

Piiper, J., and P. Scheid. 1975. Gas transfer efficacy of gills, lungs and skin: theory and experimental data. *Respir. Physiol.* 23: 209-221.

Piiper, J., and P. Scheid. 1992. Modeling of gas exchange in vertebrate lungs, gills, and skin, pp. 69-95. In: S.C. Wood, R.E. Weber, A.R. Hargens, and R.W. Millard [eds.]. Physiological Adaptations in Vertebrates: Respiration, Circulation, and Metabolism. Marcel Dekker Inc., NY, USA.

Prigogine, I., and I. Stengers. 1984. Order of Chaos: Man's New Dialogue with Nature. Heineman, London.

Romer, A.S. 1967. Major steps in vertebrate evolution. *Science* 158: 1629-1637.

Runnegar, B. 1992. Evolution of the earliest animals, pp. 65-95. In: J.W. Schopf [ed.]. Major Events of the History of Life. Johns and Bartlett, Boston, MA, USA.

Rutten, M.G. 1970. The history of atmospheric oxygen. *Space. Life. Sci.* 2: 5-17.

Scheid, P. 1987. Cost of breathing in water- and air-breathers, pp. 83-92. In: P. Dejours, C.R. Taylor, and E.R. Weibel [eds.]. Comparative Physiology: Life on Land and Water, Fidia Research Series, vol 9. Liviana Press, Padova, Venica.

Schmidt-Nielsen, K. 1972. How Animals Work. Cambridge University Press, Cambridge.

Schmidt-Nielsen, K. 1984. Scaling: Why is Animal Size so Important? Cambridge University Press, Cambridge.

Schopf, J.W. 1983. Earth's Earliest Biosphere: Its Origins and Evolution. Princeton University Press, Princeton, NJ, USA.

Slonim, N.B., and L.H. Hamilton. 1971. Respiratory Physiology, 2nd ed. The CV Mosby Company, St Louis, MO, USA.

Steen, J.B. 1971. Comparative Physiology of Respiratory Mechanisms. Academic Press, London.

Tenney, S.M. 1979. A synopsis of breathing mechanisms, pp. 51-106. In: S.C. Wood, and C. Lenfant [eds.]. Evolution of Respiratory Processes: A Comparative Approach. Marcel Dekker Inc., NY, USA.

Tsonis, A.A., and P.A. Tsonis. 1987. Fractals: A New Look, Biological Shape and Patterning. *Perspect. Biol. Med.* 30: 355-361.

Valentine, J.W., and E.M. Moores. 1976. Plate tectonics and history of life in the oceans, pp. 196-205. In: I. T. Wilson [ed.]. Continents adrift and Continents Aground. WH Freeman, San Fransisco, CA, USA.

van Valen, L. 1971. The history and stability of atmospheric oxygen. *Science* 171: 439-443.

Vollrath, F. 1992. Spider webs and silks. *Sci. Amer.* 266: 70-76.

Wagner, G.P. 1989. The origin of morphological characters and the biological basis of homology. *Evolution* 43: 1157-1171.

Wainright, S.A. 1988. Form and function in organisms. *Amer. Zool.* 28: 671-680.

Wainright, S.A., W.D. Biggs, J.D. Currey, and J.M. Gosline. 1976. Mechanical Design in Organisms. Wiley and Sons, NY, USA.

Wake, M.H. 1990. The evolution of integration of biological systems: An evolutionary perspective through studies on cells, tissues, and organs. *Amer. Zool.* 30: 897-906.

Weibel, E.R. 1984. The Pathway for Oxygen: Structure and Function in the Mammalian Respiratory System. Havard University Press, Cambridge, MA, USA.

Weibel, E.R, C.R. Taylor, and H. Hoppeler. 1991. The concept of symmorphosis: A testable hypothesis of structure-function relationship. *Proc. Natl. Acad. Sci.* 88: 10357-10361.

Weibel, E.R., C.R. Taylor, and L. Bolis. [eds.]. 1998. Principles of Animal Design: The Optimization and Symmorphosis Debate. Cambridge University Press, Cambridge.

West, B.J., and A.L. Goldberger. 1987. Physiology in fractal dimensions. *Amer. Sci.* 75: 354-365.

West, J.B. [ed.]. 1977. Bioengineering Aspects of the Lung. Marcel Dekker, NY, USA.

White, F.N. 1978. Comparative aspects of vertebrate cardiorespiratory physiology. *Ann. Rev. Physiol.* 40: 471-499.

Witt, R., and C.P. Lieckfeld. 1991. [eds.]. 1991. Bionics: Nature's Patents. Pro Futura Verlag, Munich.

Wood, S.C., and C. Lenfant [eds.]. 1979. Evolution of Respiratory Processes: A Comparative Approach. Marcel Dekker Inc, NY, USA.

2   GILLS

Adolph, E. F. 1943. On the appearance of vascular filaments on the pectoral fin of *Lepidosiren paradoxa*. *Anat. Anz.* 33: 27-30.

Alexander, R. McN. 1967. Functional Design in Fishes. Hutchison, London.

Bakhle, Y.S. 1975. Pharmacokinetic function of the lung, pp. 293-299. In: A.F. Junod, and R. Haller [eds.]. Lung Metabolism. Academic Press, NY, USA.

Baumgarten-Schumann, D., and J. Piiper. 1968. Gas exchange in the gills of resting unanaesthetised dogfish, *Scyliorhinus*. *Respir. Physiol.* 5: 317-325.

Bettex-Galland, M., and G.M. Hughes. 1973. Contractile filamentous material in the pillar cells of fish gills. *J. Cell. Sci.* 13: 359-366.

Bond, A.N. 1960. An analysis of the response of salamander gills to changes in the oxygen concentration of the medium. *Dev. Biol.* 2: 1-20.

Booth, J.H. 1978. The distribution of blood flow in the gills of fish: application of a new technique to rainbow trout (*Salmo gairdneri*). *J. Exper. Biol.* 73: 119-129.

Booth, J.H. 1979. The effect of oxygen supply, epinephrine and acetycholine on the distribution of blood flow in trout gills. *J. Exper. Biol.* 83: 31-39.

Boutilier, R.G., T.A. Heming, and G.K. Iwama. 1984. Physicochemical parameters for use in fish respiratory physiology, pp. 403-456. In: W. S. Hoar, and D. J. Randall [eds.]. Fish Physiology, vol 10A. Academic Press, NY, USA.

Brett, J.R. 1972. The metabolic demand for oxygen in fish, particularly salmonids and a comparison with other vertebrates. *Respir. Physiol.* 14: 151-174.

Burggren, W.W., and M. Doyle. 1986. Ontogeny of regulation of gill and lung ventilation in the bullfrog, *Rana catesbeiana. Respir. Physiol.* 66: 279-291.

Burggren, W.W., J. Dunn, and K. Barnard. 1979. Branchial circulation and gill morphometrics in the sturgeon, *Acipenser transmontanus. Can. J. Zool.* 57: 2160-2170.

Burggren, W.W., K. Johansen, B. McMahon. 1985b. Respiration in phyletically ancient fish, pp. 217-252. In: R. E. Foreman, A. Gorbman, J. M. Dodd, R. Olson [eds.]. Evolutionary Biology of Primitive Fishes. Plenum Press, NY, USA.

Butler, P.J., and J. D. Metcalfe. 1983. Control of respiration and circulation, pp. 41-65. In: J. C. Rankin, T. J. Pitcher, and R. Duggan [eds.]. Control Processes in Fish Physiology. Croom Helm, Beckenham, London.

Cameron, J.N., and J.J. Cech. 1970. Notes on the energy cost of gill ventilation in teleosts. *Comp. Biochem. Physiol.* 34: 447-455.

Cameron, J.N., D.J. Randall, and J.C. Davis. 1977. Regulation of the ventilation-perfusion ratio in gills of *Dasyatis sabina* and *Squalus suckleyi. Comp. Biochem. Physiol.* 39A: 505-519.

Cloutier, R., and P.L. Forey. 1991. Diversity of extinct and living actinistian fishes (Sarcopterygii). *Environm. Biol. Fishes* 32: 59-74.

Cunningham, J.T., and D.M. Reid. 1932. Experimental researches on the emission of oxygen by the pelvic filaments of the male *Lepidosiren* with some experiments on *Synbranchus marmoratus. Proc. R. Soc. Lond.* 110B: 234-248.

Dejours, P. 1973. Problems of control of breathing in fishes, pp. 117-133. In: L., Bolis, K. Schmidt-Nielsen, and S.H.P. Madrell [eds.]. Comparative Physiology. Elsevier/North Holland, Amsterdam.

De Vries R., and De Jager. 1984. The gill in the spiny dogfish, *Squalus acanthias:* respiratory and nonrespiratory function. *Amer. J. Anat.* 169: 1-29.

Dunel-Erb, S., and P. Laurent. 1980. Ultrastructure of marine teleost gill epithelia: SEM and TEM study of the chloride cell apical membrane. *J. Morph.* 165: 175-186.

Dunel-Erb, S., and P. Laurent. 1980. Functional organization of the gill vasculature in different classes of fish, pp. 37-58. In: B. Lahlou [ed.]. Epithelial Transport in the Lower Vertebrates. Cambridge University Press, Cambridge.

Emery, S.H., and A. Szczepanski. 1986. Gill dimensions in pelagic elasmobranch fishes. *Biol. Bull.* 171: 441-449.

Farrell, A.P., S.S. Sobin, D.J. Randall, and S. Crosby. 1980. Sheet blood flow in the secondary lamellae of teleost gills. *Amer. J. Physiol.* 239: R428-436.

Gannon, B.J., G. Campbell, and D.J. Randall. 1973. Scanning electron microscopy of vascular casts for the study of vessel connections in a complex vascular bed—the trout gill. Proc. Elect. Microsc. Soc. Amer. 31: 442-443.

Gilbert, C.R. 1993. Evolution and phylogeny, pp. 1-45. In: D. H. Evans [ed.]. The Physiology of Fishes. CRC Press, Boca Raton, FL, USA.

Goldstein, L. 1982. Gill nitrogen excretion, pp. 93-206. In: D.F, Houlihan, J.C. Rankin, and H.R. Shuttleworth [eds.]. Gills. Cambridge University Press, Cambridge.

Gonzalez, R.J., and D.G. McDonald. 1992. The relationsips between oxygen consumption and ion loss in freshwater fish. *J. Exper. Biol.* 163: 317-326.

Gray, I. E. 1954. Comparative study of the gill area of marine fish. *Biol. Bull. Mar. Biol. Lab., Woods Hole* 107: 219-225.

Gray, I.E. 1957. Comparative study of the gill area of crabs. *Biol. Bull. Mar. Biol. Lab., Woods Hole* 112: 34-42.

Heisler, N. 1989. Interactions between gas exchange, metabolism and ion transport in animals: an overview. *Can. J. Zool.* 67: 2923-2935.

Hills. B.A. 1996. Effects of gill dimensions on respiration, pp. 235-247. In: J. S. D. Munshi, and H. M. Dutta [eds.]. Fish Morphology: Horizon of New Reseach. Science Publishers, Lebanon, NH, USA.

Hills, B.A., and G.M. Hughes. 1970. A dimensional analysis of oxygen transfer in the fish gill. *Respir. Physiol.* 9: 126-140.

Hughes, G.M. 1966. Evolution between air and water, pp. 64-80. In: A.V.S. de Reuck, and R. Porter [eds.]. Development of the Lung. Churchill Ltd., London.

Hughes, G.M. 1978. Some features of gas transfer in fish. *Bull. Inst. Math. and Its Appli.* 14: 39-43.

Hughes, G.M. 1980. Functional morphology of fish gills, pp. 15-36. In: B. Lahlou [ed.]. Epithelial Transport in Lower Vertebrates. Cambridge University Press, Cambridge.

Hughes, G.M. 1984. General anatomy of the gills, pp. 1-72. In: D. J. Randall [ed.]. Fish Physiology, vol XA. Academic Press, London.

Hughes, G.M., and D.E. Wright. 1970. A comparative study of the ultrastructure of the water-blood pathways in the secondary lamellae of teleost and elasmobranch fishes-benthic forms. *Z Zellforsch* 104: 478-493.

Hughes, G.M., and M. Morgan. 1972. Structure of fish gills in relation to their respiratory function. *Bio. Rev.* 48: 419-475.

Kramer, D.L. 1983. The evolutionary ecology of respiratory mode in fishes: an analysis based on the costs of breathing. *Environ. Biol. Fish.* 9: 145-158.

Kylstra, J.A. 1968. Experiments in water-breathing. *Sci. Amer.* 219: 66-74.

Laurent, P. 1982. Structure of vertebrate gills, pp. 25-43. In: D.F. Houlihan, J.C. Rankin, and T.J. Shuttleworth [eds.]. Gills. Cambridge University Press, Cambridge.

Laurent, P. 1984. Gill internal morphology, pp. 73-183. In: W.S. Hoar, and D.J. Randall [eds.]. Fish Physiology, vol A. Academic Press, NY, USA.

Laurent, P., and S. Dunel-Erb. 1980. Morphology of gill epithelia. *Amer. J. Physiol.* 238: R147-R159.

Laurent, P., and S.F. Perry. 1991. Environmental effects on gill morphology. *Physiol. Zool.* 69: 2-25.

Low, W.P., D.J. Lane, and Y.K. Ip. 1988. A comparative study of terrestrial adaptations of the gills in three mudskippers—*Periophthalmus chrysospolos, Boleophthalmus boddaerti,* and *Periophthalmus schlosseri. Bio. Bull. Mar. Biol. Lab., Woods Hole* 175: 334-438.

Maina, J.N. 1990a. A study of the morphology of of the gills of an extreme alkalinity and hyperosmotic adapted teleost *Oreochromis alcalicus grahami* (Boulenger) with particular emphasis on the ultrastructure of the chloride cells and their modifications with water dilution. *Anat. Embryol.* 181: 83-98.

Maina, J.N. 1991. A morphometric analysis of chloride cells in the gills of the teleosts Oreochromis alcalicus and Oreochromis niloticus and a description of presumptive urea excreting cells in *Oreochromis alcalicus. J. Anat.* 175: 131-145.

Maina, J.N., S.M. Kisia, C.M. Wood, H.L. Bergman, A.B. Narahara, P. Laurent, and P.J. Walsh. 1996. The morphology and morphometry of the gills of *O. a. grahami:* an ecomorphological study. *Int. J. Salt Lakes Res.* 5: 131-156.

Malvin, G.M. 1989. Gill structure and function: amphibian larvae, pp. 121-151. In: S.C. Wood [ed.]. Comparative Pulmonary Physiology: Current Concepts, vol 39: Lung Biology in Health and Disease. Marcel Dekker, NY, USA.

McCutcheon, F.H. 1964. Organ systems in adaptation: the respiratory system, pp. 167-191. In: D.B. Dill, E.F. Adolph, and C.G. Wilber [eds.]. Handbook of Physiology, sect 4, Adaptation to the Environment. American Physiological Society, Washington, DC.

Milner, A.R. 1988. The relationships and origins of living amphibians, pp. 59-102. In: M.J. Benton [ed.]. The Phylogeny and Classification of Tetrapods, vol 1, Amphibians, Reptiles and Birds. Clarendon Press, Oxford.

Muir, B.S., and J.I. Kendall. 1968. Structural modifications in the gills of tunas and some other oceanic fishes. *Copeia* 1968: 388-398.

Neckvasil, N.P., and K.R. Olson. 1986. Extraction and metabolism of circulating catecholamines by the trout gill. *Amer. J. Physiol.* 19: R5276-5287.

Newman, R.A. 1992. Adaptive plasticity in amphibian metamorphosis. *BioScience* 42: 671-678.

Nilsson, S. 1985. Filament position in fish gills is influenced by a smooth muscle enervated by adrenergic nerves. *J. Exper. Biol.* 118: 433-437.

Nilsson, S. 1986. Control of gill blood flow, pp. 86-101. In: S. Nilsson, and S. Holmgren [eds.]. Fish Physiology: Recent Advances. Croom Helm, Dover, NH, USA.

Ojha, J., and S.K. Singh. 1986. Scanning electron microscopy of the gills of a hill-stream fish, *Danio dangila* [Ham.]. *Arch. Biol., Bruxelles* 97: 455-467.

Olson, K.R. 1991. Vasculature of the fish gills: anatomical correlates of physiological function. *J. Electron. Microsc. Tech.* 2: 217-228.

Olson, K.R., and P.O. Fromm. 1973. A scanning electron microscope study of secondary lamellae and chloride cells of rainbow trout (*Salmo gairdneri*). *Z. Zellforsch.* 143: 439-449.

Part, P., H. Tuurala, M. Nikinmaa, and A. Kiessling. 1984. Evidence for nonrespiratory intralamellar shunt in perfused rainbow trout gills. *Comp. Biochem. Physiol.* 79A: 29-34.

Perry, S.F., and G. McDonald. 1993. Gas exchange, pp. 251-278. In: D. H. Evans [ed.]. The Physiology of Fishes. CRC Press, Boca Raton.

Piiper, J., P. Scheid, S.F. Perry, and G.M. Hughes. 1986. Effective and morphometric oxygen diffusing capacity of the gills of the elasmobranch *Scyliorhinus stellaris*. *J. Exper. Biol.* 123: 27-41.

Randall, D.J., D. Baumgarten, and M. Malyusz. 1972. The relationship between gas and ion transfer across the gills of fishes. *Comp. Biochem. Physiol.* 41A: 629-637.

Satchell, G.H. 1971. Circulation in Fishes. Cambridge University Press, Cambridge.

Sperry, D.G., and R.J. Wassersug. 1976. A proposed function for microridges on epithelial cells. *Anat. Rec.* 185: 253-258.

Tota, B., and W.C. Hamlett. 1989. Epilogue: Evolutionary and contemporary biology of elasmobranchs. *J. Exper. Zool.* 2: 193-196.

Uchiyama, M., H. Yoshizawa, C. Wakasugi, and C. Oguro.1990. Structure of the internal Gills of tadpoles of the crab -eating frog, *Rana cancrivora. Biol. Sci.* 7: 623-630.

Zadunaisky, J.A. 1984. The chloride cell: the active transport of chloride and the paracellular pathways, In: W.S. Hoar, and D.J. Randall [eds.]. Fish Physiology, vol. XB. Academic Press, London.

3 SKIN

Beachy, C.K., and R.C. Bruce. 1992. Lunglessness in Plethodontid salamanders is consistent with the hypothesis of a mountain stream origin: a response to Ruben and Boucot. *Amer. Nat.* 139: 839-847.

Belkin, D.A. 1968. Aquatic respiration and underwater survival of two freshwater turtle species. *Respir. Physiol.* 4: 1-14.

Boutilier, R.G., and D.P. Toews. 1981. Respiratory, circulatory and acid-base, adjustments to hypercapnia in a strictly aquatic and predominantly skin-breathing urodele *Cryptobranchus alleganiensis. Respir. Physiol.* 46: 177-192.

Burggren, W.W., and M.E. Feder. 1985. Skin breathing in vertebrates. *Sci. Amer.* 253: 106-118.

Burggren, W.W., M.L. Glass, and K. Johansen. 1977. Pulmonary ventilation/perfusion relationships in terrestrial and aquatic chelonian reptiles. *Can. J. Zool.* 55: 2024-2034.

Crawford, E.C., and R.R. Schultetus. 1970. Cutaneous gas exchange in the lizard, *Sauromalus obesus. Copeia* 1970: 179-180.

Czopek, J. 1962. Vascularization of respiratory surfaces in some caudata. *Copeia* 1962: 576-587.

Czopek, J. 1965. Quantitative studies of the morphology of respiratory surfaces in amphibians. *Acta. Anat.* 62: 296-323.

Czopek, J., and H. Szarski. 1989. Morphological adaptations to water movements in the skin of anuran amphibians. *Acta. Biol., Cracov* 31: 81-96.

Dupré, R.K., W.W. Burggren, and T.Z. Vitalis. 1991. Dehydration decreases cutaneous gas diffusing capacity in the toad, *Bufo woodhouseii. Amer. Zool.* 31:75A.

Eastman, J.T. 1993. Antarctic Fish Biology: Evolution in a unique Environment. Academic Press, San Diego, CA, USA.

Ewer, D.W. 1959. A toad (*Xenopus laevis*) without haemoglobin. *Nature, Lond* 183: 271.

Fedde, M.R. 1980. The structure and gas-flow pattern in the avian respiratory system. *Poult. Sci.* 59: 2642-2653.

Feder, M.E., and W.W. Burggren. 1985. Cutaneous gas exchange in vertebrates: design, patterns, control, and implications. *Biol. Rev.* 60: 1-45.

Feder, M.E., and W.W. Burggren. 1985. The regulation of cutaneous gas exchange in vertebrates, pp. 101-112. In: R. Giles [ed.]. Circulation, Respiration, and Metabolism. Springer-Verlag, Berlin-Heidelberg.

Feder, M.E., and A.W. Pinder. 1988. Ventilation and its effect on "infinite pool" exchangers. *Amer. Zool.* 28: 973-983.

Full, R. J. 1985. Exercising without lungs: Energetics and endurance in a lungless salamander, *Plethodon jordani. Physiologist* 28: 342.

Gatz, R.N., E.C. Crawford, and J. Piiper. 1974. Respiratory properties of the blood of lungless and gill-less salamander, *Desmognathus fuscus. Respir. Physiol.* 20: 33-41.

Hutchison, V.H., H.B. Haines, and G. Engbretson. 1976. Aquatic life at high altitude: respiratory adaptations in the Lake Titicaca frog, *Telmatobius culeus. Respir. Physiol.* 27: 115-129.

Johansen, K., G. Lykkeboe, S. Kornerup, and G.M.O. Maloiy. 1980. Temperature insensitive oxygen binding in blood of the tree frog *Chiromantis petersi. J. Comp. Physiol.* 136: 71-76.

Kirsch, R., and G. Nonnotte. 1977. Cutaneous respiration in three fresh water teleosts. *Respir. Physiol.* 29: 339-354.

Lenfant, C., and K. Johansen.1967. Respiratory adaptations in selected amphibians. *Respir. Physiol.* 2: 247-260.

Lutcavage, M.E., P.L. Lutz, and H. Baier. 1987. Gas exchange in the loggerhead sea turtle, *Caretta. J. Exper. Biol.* 131: 365-372.

Malvin, G.M. 1988. Microvascular regulation of cutaneous gas exchange in amphibians. *Amer. Zool.* 28: 999-1007.

Malvin, G.M., and M.P. Hlastala. 1986. Regulation of cutaneous gas exchange by environmental O_2 and CO_2 in the frog. *Respir. Physiol.* 65: 99-111.

Malvin, G.M., and N. Heisler. 1988. Blood flow patterns in the salamander, *Ambryostoma tigrinum*, before, during and after metamorphosis. *J. exp. Biol.* 137: 53-74.

Moalli, R., R.S. Meyers, D.C. Jackson, and R.W. Millard. 1980. Skin circulation of the frog, *Rana catesbeiana:* distribution and dynamics. *Respir. Physiol.* 40: 137-148.

Noble, G.K. 1925. Integumentary, pulmonary, and cardiac modifications correlated with increased cutaneous respiration in the amphibian: a solution to the "hairy frog" problem. *J. Morph. Physiol.* 40: 341-416.

Noble, G.K. 1929. The adaptive modifications of the arboreal tadpoles of *Hoplophryne* and torrent tadpoles of *Staurois*. *Bull. Amer. Mus. Nat. Hist.* 58: 291-334.

Noble, G.K. 1931. The Biology of Amphibia. McGraw-Hill, NY, USA.

Piiper, J., R.N. Gatz, and E.C. Crawford. 1976. Gas transport characteristics in an exclusively skin breathing salamander, *Desmognathus fuscus* (Plethodontidae), pp. 339-356. Respiration in Amphibious Vertebrates. Academic Press, London.

Romer, A.S. 1972. Skin breathing-primary or secondary? *Respir. Physiol.* 14: 183-192.

Ruben, J.A., N.L. Reagan, P.A. Verrell, and A.J. Boucot. 1993. Plethodontid salamander origins: a response to Beachy and Bruce. *Amer. Nat.* 142:1038-1051.

Standaert, T., and K. Johnsen. 1974. Cutaneous gas exchange in snakes. *J. Comp. Physiol.* 89: 313-320.

4 SWIM (AIR) BLADDER

Alexander, R. McN. 1993. Buoyancy, pp. 75-97. In: D.H. Evans [ed.]. The Physiology of Fishes. CRC Press, Boca Raton, FL, USA.

Brooks, R. E. 1970. Ultrastructure of the physostomtous swim bladder of rainbow trout *Salmo gairdneri*. *Z. Zellforsch.* 106: 473- 483.

Copeland, D.E. 1969. Fine structural study of gas secretion in the physoclistous swim bladder of *Fundulus heteroclitus* and *Gadus callarias* and in the euphysoclistous swim bladder of *Opsanus tau*. *Z. Zellforsch* 93: 305-331.

D'Aoust, G. 1970. The role of lactic acid in gas secretion in the teleost swim bladder. *Comp. Biochem. Physiol.* 32: 637-668.

Dehadrai, P.V. 1962. Respiratory function of the swim bladder of *Notopterus* (Lacepede). *Nature, Lond.* 185: 929.

Dickson, K.A., and J.B. Graham. 1986. Adaptations to hypoxic environments in the erythrinid fish, *Hoplias microlepis*. *Environ. Biol. Fishes* 15: 301-308.

Evans, H.M. 1929. Some notes on the anatomy of the electric eel, *Gymnotus electrophorus*, with special reference to a mouth-breathing organ and the swim bladder. *Proc. Zool. Soc. Lond* 57: 17-23.

Evans, H.M., and G.C.C. Damant. 1928. Observations on the physiology of the swim bladder of cyprinoid fishes. *J. Exper. Biol.* 6: 42-55.

Fahlén, G. 1971. The functional morphology of the gas bladder of the genus *Salmo*. *Acta. Anat.* 78: 161-184.

Fänge, R. 1983. Gas exchange in the fish swim bladder. *Rev. Physiol. Biochem. Pharmacol.* 97: 111-158.

Gee, J.H. 1976. Buoyancy and aerial respiration: factors influencing the evolution of reduced swim bladder volume of some Central American catfishes (Trichomycteridae, Callichthyidae, Loricariidae, Astroblepidae). *Can. J. Zool.* 54: 1030-1037.

Gee, J.H. 1981. Coordination of respiratory and hydrostatic functions of the swim bladder in the Central American mudminnow, *Umbra limi*. *J. Exper. Biol.* 92: 37-53.

Gee, J.H., and P.A. Gee. 1995. Aquatic surface respiration, buoyancy control and the evolution of air-breathing gobies (Gobiidae: Pisces). *J. Exper. Biol.* 198: 79-89.

Gerth, W.A., and F.A. Hemmingsen. 1982. Limits of gas secretion by salting-out effect in the fish swim bladder rete. *J. Comp. Physiol.* 146B: 129-136.

Graham, J.B., J.H. Gee, and F.S. Robinson. 1975. Hydrostatic and gas exchange functions of the lung of the sea snake, *Pelamis platurus*. *Comp. Biochem. Physiol.* 50: 477-482.

Graham, J.B., R.H. Rosenblatt, and C. Gans. 1978. Vertebrate air breathing arose in fresh waters and not in the ocean. *Evolution* 32: 459-463.

Guimond, R.W., and V.H. Hutchison. 1976. Gas exchange of the giant salamanders of North America, pp. 313-338. In: G. M. Hughes [ed.]. Respiration of Amphibious Vertebrates. Academic Press, London.

Kanwisher, J.W., and A. Ebeling. 1957. Composition of the swim bladder gas in bathypelagic fishes. *Deep Sea Res.* 2: 211-223.

Kuhn, W., A. Ramel, H. Kuhn, and E. Marti. 1963. The filling mechanism of the swim bladder. Generation of high gas pressures through hairpin counter-current multiplication. *Experientia* 19: 497-511.

Lapennas, G.N., and K. Schimdt-Nielsen. 1977. Swim bladder permeability to oxygen. *J. Exper. Biol.* 67: 175-196.

Liem, K.F. 1989. Respiratory gas bladders in teleosts: Functional conservatism and Morphological diversity. *Amer. Zool* 29: 333-352.

Marshall, N.B. 1960. Swim bladder structure of deep sea fishes in relation to their systematics and biology. Discovery Report 31: 1-122.

Pelster, B., and P. Scheid. 1991. Activities of enzymes for glucose catabolism in the swim bladder of the European eel *Anguilla anguilla*. *J. Exper. Biol.* 156: 207-213.

Pelster, B., and P. Scheid. 1992. Counter current concentration and gas secretion in the fish swim bladder. *Physiol. Zool* 65:1-16.

Pelster, B., and P. Scheid. 1992. Metabolism of the swim bladder epithelium and the single concentrating effect. *Comp. Biochem. Physiol.* 105A: 383-388.

Pelster, B., and P. Scheid. 1992. The influence of gas gland metabolism and blood flow on gas deposition into the swim bladder of the European eel *Anguilla anguilla*. *J. Exper. Biol.* 173: 205-216.

Pelster, B, and P. Scheid. 1993. Glucose metabolism of the swim bladder tissue of the European eel, *Anguilla anguilla*. *J. Exper. Biol.* 185: 169-178.

Pelster, B., H. Kobayashi, and P. Scheid. 1989. Metabolism of the perfused swim bladder of the European eel: oxygen, carbon dioxide, glucose and lactate balance. *J. Exper. Biol.* 144: 495-506.

Phleger, C.F., and B.S. Saunders. 1978. Swim bladder surfactants of Amazon air breathing fishes. *Can. J. Zool* 56: 946-952.

Piiper, J., H.T. Humphrey, and H. Rahn. 1962. Gas composition of pressurized, perfused gas pockets and the fish swim bladder. *J. Appl. Physiol.* 17: 275-282.

Potter, G.E. 1927. Respiratory function of swim bladder in *Lepidosteus. J. Exper. Zool.* 49: 45-52.

Saunders, R.L. 1953. The swim bladder gas content of some fresh water fish with particular reference to the physostomes. *Can. J. Zool.* 31: 547-560.

Scholander, P.F. 1954. Secretion of gases against high pressures in the swim bladder of deep sea fishes. II. The rete mirabile. *Biol. Bull. Mar. Biol. Lab., Woods Hole* 107: 247-259.

Scholander, P.F., L. van Dam. 1953. Composition of the swim bladder gas in deep sea fishes. *Biol. Bull.* 104: 75-97.

Todd, E.S. 1973. Positive buoyancy and air-breathing: A new piscine gas bladder function. *Copeia* (1973): 461-464.

Wittenberg, J.B. 1965. The secretion of oxygen into the swim bladder of fish. *J. Gen. Physiol.* 44: 521-526.

5 TRANSITIONAL (BIMODAL) BREATHING

Barrell, J. 1916. Influence of Silurian-Devonian climates on the rise of air breathing vertebrates. *Bull. Geol. Soc. Am.* 27: 387-436.

Barton, M., and K. Elkins. 1988. Significance of aquatic surface respiration in the comparative adaptation of two species of fishes (*Notropis chrysocephalus* and *Fundulus catenus*) to headwater environments. *Trans. KY. Acad. Sci.* 49: 69-73.

Beitinger, T.L., and M.J. Pettit 1984. Comparison of low oxygen avoidance in a bimodal breather, Erpetoichthys calabaricus and an obligate water breather, *Percina caprodes. Environ. Biol. Fishes* 11: 235-240.

Ben-Avraham, Z. 1981. The movement of the continents. *Amer. Sci.* 69: 291-299.

Bennett, M.B. 1988. Morphometric analysis of the gills of the European eel, *Anguilla anguilla. J. Zool. Lond.* 215: 549-560.

Berg, T., and J.B. Steen. 1965. Physiological mechanisms for aerial respiration in the eel. *Comp. Biochem. Physiol.* 15: 469-484.

Boutilier, R.G. 1990. Respiratory gas tensions in the environment, pp. 1-13. In: R.G. Boutilier [ed.]. Advances in Comparative and Environmental Physiology, vol 6: Vertebrate Mass Exchange from Environment to Cell. Springer-Verlag, Berlin–Heidelberg.

Boutilier, R.G., M.L. Glass, and N. Heisler. 1986. The relative distribution of pulmocutaneous blood flow in *Rana catesbeiana:* effects of pulmonary or cutaneous hypoxia. *J. Exper. Biol.* 126: 33-39.

Brainerd, E.L. 1994. Lung ventilation in fishes and amphibians: the evolution of vertebrate air-breahing mechanisms. *Amer. Zool.* 34: 289-299.

Browman, M.W., and D.L. Kramer. 1985. *Pangasius sutchi* (Pangassidae), an air-breathing catfish that uses the swim bladder as an accessory respiratory organ. *Copeia* 1985: 994-998.

Budgett, J.S. 1900. Observations on *Polypterus* and *Protopterus. Proc. Cambr. Philos. Soc.* 10: 236-240.

Burggren, W.W. 1979. Bimodal gas exchange during variation in environmental oxygen and carbon dioxide in the air-breathing fish *Trichogaster trichopterus. J. Exper. Biol.* 82: 97-213.

Burggren, W.W., and R.L. Infantino. 1994. The respiratory transition from water to air breathing during amphibian metamorphosis. *Amer. Zool.* 34: 238-246.

Burggren, W.W., and K. Johansen. 1986. Circulation and respiration in lungfishes (Dipnoi). *J. Morph. Suppl.* 1: 217-236.

Burggren, W.W ., and K. Johansen. 1987. Circulation and respiration in lungfishes (Dipnoi) pp. 217-236. In: W.E. Bemis, W.W. Burggren, and N.E. Kemp [eds.]. The Biology and Evolution of Lungfishes. Alan R. Liss, NY, USA.

Burggren, W.W., and N.H. West. 1982. Changing importance of gills, lungs and skin during metamorphosis in the bullfrog *Rana catesbeiana. Respir. Physiol.* 47: 151-164.

Carter, G.S. 1957. Air breathing, pp. 65-79. In: M.E. Brown [ed.]. The Physiology of Fishes. Academic Press, London.

Carter, G.S., and L.C. Beadle. 1930. Notes on the habits and development of *Lepidosiren paradoxa. J. Linn. Soc. Zool.* 37: 327-368.

Carter, G.S., and L.C. Beadle. 1931. The fauna of the swamps of the Paraguayan Chaco in relation to its environment: respiratory adaptations in fishes. *J. Linn. Soc., Zool.* 37: 205-258.

Das, B.K. 1927. The bionomics of certain air breathing fishes of India together with an account of the development of the air breathing organs. *Phil. Trans. R. Soc., Lond.* 216B: 183-219.

Das, B.K. 1940. Nature and causes of evolution and adaptation of the air breathing fishes. Proc 27th Indian Sci. Congr. 2: 215-260.

Davenport, J. 1985. Environmental Stress and Behavioural Adaptation. Croom Helm, London.

Dehadrai, P.V., and S.D. Tripathi. 1976. Environment and ecology of fresh water air-breathing teleosts, pp. 39-72. In: G.M. Hughes [ed.]. Respiration of Amphibious Vertebrates. Academic Press, London.

Dejours, P. 1994. Environmental factors as determinats in bimodal breathing: an introductory overview. *Amer. Zool.* 34: 178-183.

DeLaney, R.G., and A.P. Fishman. 1977. Analysis of lung ventilation in the aestivating lungfish *Protopterus aethiopicus. Amer. J. Physiol.* 233: R181-187.

DeLaney, R.G., G.S. Lahiri, and A.P. Fishman. 1974. Aestivation of the African lungfish *Protopterus aethiopicus:* cardiovascular and respiratory functions. *J. Exper. Biol.* 61: 111-118.

Driedzic, W.R., C.F. Phleger, S.H. Fields, and C. French. 1978. Alterations in energy metabolism associated with transitions from water to air breathing fish. *Can. J. Zool.* 56: 730-735.

Dubale, M.S. 1959. A comparative study of the oxygen carrying capacity of the blood in water and air-breathing teleosts. *J. Anim. Morphol.* Physiol. 6: 48-54.

Faber, J.J., and H. Rahn. 1970. Gas exchange between air and water and the ventilation pattern of the electric eel. *Respir. Physiol.* 9: 151-161.

Farrell, A.P., and D.J. Randall. 1978. Air breathing mechanisms in two Amazonian teleosts, *Arapaima gigas* and *Hoplerythrinus unitaeniatus. Can. J. Zool.* 56: 939-945.

Gahlenbeck, H., and H. Bartels. 1970. Blood gas transport properties in gill and lung forms of the axolotol (*Ambryostoma mexicanum*). *Respir. Physiol.* 9: 175-182.

Gannon, B.J., D.J. Randall, J. Browning, R.J.G. Lester, and L.J. Rogers. 1983. The microvascular organization of the gas exchange organs of the Australian lungfish, *Neoceratodus forsteri* (Krefft). *Aust. J. Zool.* 31: 651-673.

Glass, M.L., A. Ishmatsu, and K. Johansen. 1986. Responses of aerial ventilation to hypoxia and hypercapnia in *Channa argus,* an air-breathing fish. *J. Comp. Physiol.* B156: 425-430.

Gordon, M.S.,W.W. Ng, A.Y. Yip. 1978. Aspects of terrestrial life in amphibious fishes. III. The Chinese mudskipper, *Periophthalmus cantonensis. J. Exper. Biol.* 72: 57-75.

Gordon, M.S., I. Boetius, D.H. Evans, R. McCarthey, and L.C. Oglesby. 1969. Aspects of the physiology of terrestrial life in amphibious fishes. I. The mudskipper *Periophthalmus sobrinus. J. Exper. Biol.* 50: 141-149.

Gorr, T., T. Kleinschmidt, and H. Fricke. 1991. Close tetrapod relationships of the coelacanth *Latimeria* indicated by haemoglobin sequences. *Nature, Lond.* 351: 394-397.

Graham, J.B. 1976. Respiratory adaptations of marine air-breathing fishes, pp. 165-187. In: G.M. Hughes [ed.]. Respiration of Amphibious Vertebrates. Academic Press, London.

Graham, J.B. 1990. Ecological, evolutionary, and physical factors influencing aquatic animal respiration. *Amer. Zool.* 30: 137-146.

Graham, J.B. 1994. Air Breathing Fishes: Evolution, Diversity, and Adaptation. Academic Press, San Diego, CA, USA.

Graham, J.B. 1994. An evolutionary perspective for bimodal respiration: a biological synthesis of fish air breathing. *Amer. Zool.* 34: 229-237.

Graham, J.B., D.L. Kramer, and E. Pineda. 1978. Comparative respiration of an air breathing and non-air breathing characoid fish and the evolution of aerial respiration in characins. *Physiol. Zool.* 51: 279-288.

Graham, J.B., R. Dudley, N.M. Agullar, and C. Gans. 1995. Implications of the late Paleozoic oxygen pulse fo physiology and evolution. *Nature, Lond.* 375: 117-120.

Guimond, R.W., and H.V. Hutchison. 1972. Pulmonary branchial and cutaneous gas exchange in the mudpuppy, *Necturus maculosus maculosus* (Rafinesque). *Comp. Biochem. Physiol.* 42A: 367-393.

Hakim, A., J.S.D. Munshi, and G.M. Hughes. 1978. Morphometrics of the respiratory organs of the Indian green snakehead fish, *Channa punctata. J. Zool., Lond.* 184: 519-543.

Hochachka, P.W., J. Fields, and T. Mustafa. 1973. Animal life without oxygen: basic biochemical mechanisms. *Amer. Zool.* 13: 543-555.

Howell, B.J. 1970. Acid-base balance in transition from water to air breathing. *Fed. Proc.* 29: 1130-1134.

Hughes, G.M. 1966. Evolution between air and water, pp. 64-80. In: A.V.S. de Reuck, and R. Porter [eds.]. Development of the Lung. Churchill Ltd., London.

Hughes, G.M. 1976. [ed.]. Respiration of Amphibious Vertebrates. Academic Press, London.

Hughes, G.M. 1995. The gills of the coelacanth, *Latimeria chalumnae;* a study in relation to body size. Phil. *Trans. R. Soc., Lond.* 347B: 427-438.

Hughes, G.M., and M. Morgan. 1973. The structure of fish gills in relation to their respiratory function. *Biol. Rev.* 48: 419-475.

Hughes, G.M., and E.R. Weibel. 1976. Morphometry of fish lungs, pp. 213-232. In: G.M. Hughes [ed.]. Respiration of Amphibious Vertebrates. Academic Press, London.

Hughes, G.M., and J.S.D. Munshi. 1979. Fine structure of the gills of some Indian air-breathing fishes. *J. Morph.* 160:169-194.

Ishmatsu, A., and Y. Itazawa. 1981. Ventilation of the air breathing organ in the snakehead *Channa argus. Jap. J. Ichthyol.* 28: 276-282.

Jasinski, A. 1973. Air-blood barrier in the respiratory intestine of the pond roach, *Misgurnus fossilis* L *Acta. Anat.* 86: 376-391.

Jesse, M.J., C. Shub, and A.P. Fishman. 1967. Lung and gill ventilation of the African lungfish. *Respir. Physiol.* 3: 267-287.

Johansen, K. 1970. Air-breathing in fish, pp. 361-411. In: W.S. Hoar, and D.J. Randall [eds.]. Fish Physiology, vol 4. Academic Press, London.

Johansen, K., and C. Lenfant. 1967. Respiratory function in the South American lungfish. *J. Exper. Biol.* 46: 205-218.

Johansen, K., and C. Lenfant. 1968. Respiration in the African lungfish, *Protopterus aethiopicus.* II. Control of breathing. *J. Exper. Biol.* 49: 453-468.

Johansen, K., C. Lenfant, and G.F. Grigg. 1967. Respiratory control in the lungfish, *Neoceratodus forsteri* (Krefft). *Comp. Biochem. Physiol.* 20: 835-854.

Just, J.J., R.N. Gatz, and E.C. Crawford. 1973. Changes in respiratory functions during metamorphosis of the bullfrog, *Rana catesbeiana. Respir. Physiol.* 17: 276-282.

Kimura, A., T. Gomi, Y. Kikuchi, and T. Hashimoto. 1987. Anatomical studies of the lung of air breathing fish. I. Gross anatomical and light microscopic observations of the lungs of the African lungfish *Protopterus aethiopicus. J. Med. Soc. (Toho) Japan* 34: 1-18.

Klika, E., and A. Lelek. 1967. A contribution to the study of the lungs of *Protopterus annectens* and *Polypterus senegalensis. Folia Morph.* 15: 168-175.

Kramer, D.L., and M. McClure. 1982. Aquatic surface respiration, widespread adaption to hypoxia in tropical fresh water fishes. *Environ. Biol. Fishes* 7: 47-55.

Lahiri, S., J.P. Szidon, and A.P. Fishman. 1970. Potential respiratory and circulatory adjustments to hypoxia in the African lungfish. *Fed. Proc.* 29: 1141-1148.

Laurent, P. 1996. Vascular organization of lungfish, a landmark in ontogeny and phylogeny of air-breathers, pp. 47-58. In: J.S.D. Munshi, and H.M. Dutta [eds.]. Fish morphology: Horizon of New Research. Science Publishers Inc., Lebanon, NH, USA.

Laurent, P., R.G. Delaney, and A.P. Fishman. 1978. The vasculature of the gills in the aquatic and aestivating lungfish (*Protopterus aethiopicus*). *J. Morph.* 156: 173-208.

Lenfant, C., and K. Johansen. 1968. Respiration in an African lungfish, *Protopterus aethiopicus:* respiratory properties of blood and normal patterns of breathing and gas exchange. *J. Exper. Biol.* 49: 437-452.

Lenfant, C., and K. Johansen. 1972. Gas exchange in gill, skin and lung breathing. *Respir. Physiol.* 14: 211-218.

Lenfant, C., K. Johansen, and D. Hanson. 1970b. Bimodal gas exchange and ventilation-perfusion relationship in lower vertebrates. *Fed. Proc.* 29: 1124-1129.

Liem, K.F. 1981. Larvae of air-breathing fishes as counter-current flow devices in hypoxic environments. *Science* 211: 1177-1179.

Liem, K.F. 1987. Form and function of lungs: The evolution of air breathing mechanisms. *Amer. Zool.* 28: 739-759.

Liem, K.F. 1987. Functional design of the air ventilation apparatus and overland excursions by teleosts. *Fieldiana. Zool.*, New Ser 37: 1-29.

Little, C. 1990. The terrestrial invasion: An ecophysiological approach to the origins of land animals. Cambridge University Press, Cambridge.

Lomholt, J.P., and K. Johansen. 1976. Gas exchange in the amphibious fish, *Amphipnous cuchia. J. Comp. Physiol.* 107: 141-157.

Lomholt, J.P., K. Johansen, and G.M.O. Maloiy. 1975. Is aestivating lungfish the first vertebrate with suctional breathing? *Nature, Lond.* 257: 787-788.

Maina, J.N. 1987. The morphology of the lung of the African lungfish, *Protopterus aethiopicus:* a scanning electron microscopic study. *Cell Tissue Res.* 250: 191-196.

Maina, J.N. 1994. Comparative pulmonary morphology and morphometry: the functional design of respiratory systems, pp. 111-232. In: R. Gilles [ed.]. Advances in Comparative and Environmental Physiology, vol 20. Springer-Verlag, Heidelberg.

Maina, J.N. 1998. The Gas Exchangers: Structure, Function and Evolution of the Respiratory Processes. Springer-Verlag, Heidelberg.

Maina, J.N. 2000. Comparative respiratory morphology: themes and principles in the design and construction of the gas exchangers. *Anat. Rec.* 261: 25-44.

Maina, J.N., and G.M.O. Maloiy. 1985. The morphometry of the lung of the lungfish (*Protopterus aethiopicus*): its structural-functional correlations. *Proc. R. Soc. Lond.* 244B: 399-420.

Maina, J.N., and G.M.O. Maloiy. 1986. The morphology of the respiratory organs of the African air-breathing catfish (*Clarias mossambicus*): A light, and electron microscopic study, with morphometric observations. *J. Zool., Lond.* 209: 421-445.

McMahon, B.R. 1969. A functional analysis of the aquatic and aerial respiratory movements of an African lungfish, *Protopterus aethiopicus,* with refrence to the evolution of the lung ventilation mechanism in vertebrates. *J. Exper. Biol.* 51: 407-430.

Moore, J.A. 1990b. The ability to live on dry land, rather than in water, required major adjustments in structure and physiology. *Amer. Zool.* 30: 847-849.

Morris, S., and C.R. Bridges. 1994. Properties of respiratory pigments in bimodal breathing animals: air and water breathing by fish and crustaceans. *Amer. Zool.* 34: 216-228.

Munshi, J.S.D. 1976. Gross and fine structure of the respiratory organs of air-breathing fishes, pp. 73-102. In: G.M. Hughes [ed.]. Respiration of Amphibious Vertebrates. Academic Press, London.

Munshi, J.S.D., and G.M. Hughes. 1992. Air Breathing Fishes of India: Their Structure, Function and Life History. AA Balkema Uitgevers BV, Rotterdam.

Olson, K.R. 1994. Circulation in bimodally breathing fish. *Amer. Zool.* 34: 280-288.

Packard, G.C. 1974. The evolution of air-breathing in Paleozoic gnathostome fishes. *Evolution* 28: 320-325.

Pelzenberger, M., and H. Pohla. 1992. Gill surface area of water and air breathing fish. *Rev. Fish Biol. Fisheries* 2: 187-216.

Peters, H.M. 1978. On the mechanism of air ventilation in anabantoids (Pisces, Teleostei). *Zoomorphologie* 89: 93-123.

Prasad, M.S. 1988. Morphometrics of gills during growth and development of air- breathing habit in *Colisa fasciatus* (Bloch and Schneider). *J. Fish. Biol.* 32: 367-381.

Rahn, H., and B.J. Howell. 1976. Bimodal gas exchange, pp. 271-285. In: G.M. Hughes [ed.]. Respiration of Amphibious Vertebrates. Academic Press, London.

Randall, D.J., W.W. Burggren, A.P. Farrell, and M.S. Haswell. 1981. The Evolution of Air-breathing in Vertebrates. Cambridge University Press, Cambridge.

Rosen, D.E., P.L. Forey, B.G. Gardiner, and C. Petterson. 1981. Lungfishes, tetrapods paleontology and plesiomorphology. *Bull. Am. Mus. Nat. Hist.* 167: 159-276.

Sacca, R., and W.W. Burggren. 1982. Oxygen uptake in water and air in the air-breathing reedfish Calamoichthys calabaricus role of skin, gills and lungs. *J. Exper. Biol.* 97: 179-186.

Satchell, G.H. 1976. The circulatory system of air-breathing fish, pp. 105-124. In: G.M. Hughes [ed.]. Respiration of Amphibious Vertebrates. Academic Press, London.

Saxena, D.B. 1963. A review of ecological studies and their importance in the physiology of air breathing fish. *Ichthyologica* 2: 116-128.

Sayer, M.D.J., and J. Davenport. 1991. Amphibious fish: Why do they leave water? *Rev. Fish Biol. Fisheries* 1: 159-181.

Selden, P., and D. Edwards. 1989. Colonization of the land, pp. 67-127. In: K.C. Allen, and D.E.G. Briggs [eds.]. Evolution and Ecology. Pinter, London.

Shelton, G., and P.C. Croghan. 1988. Gas exchange and its control in non-steady state systems: the consequences of evolution from water to air breathing in vertebrates. *Can. J. Zool.* 66: 109-123.

Singh, B.N., and G.M. Hughes. 1973. Respiration of an air-breathing catfish, *Clarias batrachus* (Linn). *J. Exper. Biol.* 55: 421-434.

Stevens, E.D., and G.F. Holeton. 1978. The partitioning of oxygen uptake from air and from water by the large obligate air breathing teleost, pirarucu (*Arapaima gigas*). *Can. J. Zool.* 56: 974-976.

Ultsch, G.R. 1976. Eco-physiological studies of some metabolic and respiratory adaptations of sirenid salamanders, pp. 278-312. In: G.M. Hughes [ed.]. Respiration of Amphibious Vertebrates. Academic Press, London.

Val, A.L., V.M. Fonseca, and F.G. Affonso. 1990. Adaptive features of Amazon fishes: haemoglobins, hematology, intraerythrocytic phosphates and whole blood Bohr effect of *Pterygoplichthys multiradiatus* (Siluriformes). *Comp. Biochem. Physiol.* 97B: 435-440.

Weber, R.E., S.C. Wood, and B.J. Davis. 1979. Acclimation to hypoxic water in facultative air-breathing fish: blood oxygen affinity and allosteric effectors. *Comp. Biochem. Physiol.* 62A: 125-129.

Wu, E.R. 1993. The development and evolution of a key morphological innovation: air-breathing organs in the anabantoidei. *Amer. Zool.* 33: 14A .

Wu, H.W., and H.W. Chang. 1947. On the arterial system of the gills and the suprabranchial cavities in *Ophiocephalus argus*, with special reference to the correlation with bionomics of the fish. *Sinensia* 17: 1-15.

Zaccone, G., S. Fasulo, and L. Ainis. 1995. Gross anatomy, histology and immunohistochemistry of respiratory organs of air-breathing and teleost fishes, pp 15-33. In: L.M. Pastor [ed.]. Histology, Ultrastructure and Immunohistochemistry of Respiratory Organs in Non-mammalian Vertebrates. Servicio de Publicaciones de la Universidad de Murcia, Murcia, Spain.

6   AMPHIBIAN LUNG

Bell, P.B., and V.I. Stark-Vancs. 1983. SEM study of the microarchitecture of the lung of the giant salamnder *Amphiuma tridactylum*. *Scan. Electr. Microsc.* 1983: 449-456.

Bennett, A.F. 1978. Activity metabolism in the lower vertebrates. *Ann. Rev. Physiol.* 400: 447-469.

Bennett, A.F., and M.H. Wake. 1974. Metabolic correlates of activity in the caecilian, *Geotrypetes seraphini*. *Copeia* 1874: 764-769.

Bentley, P.J. and J.W. Shield. 1973. Respiration of some urodele and anuran amphibia. II. In air, role of the skin and lungs. *Com. Biochem. Physiol.* 46A: 29-38.

Black, C.P., S.M. Tenney, and M.V. Kroonenburg. 1978. Oxygen transport during progressive hypoxia in bar-headed geese (*Anser anser*) acclimated to sea level and 5600 m, pp. 79-83. In: J. Piiper [ed.]. Respiratory Function in Birds Adult and Embryonic. Springer-Verlag, Berlin.

Brainerd, E.L., J.S. Ditelberg, and D.M. Bramble. 1993. Lung ventilation in salamanders and the evolution of vertebrate air breathing mechanisms. *Biol. J Linn. Soc.* 49: 163-183.

Broyles, R.H. 1981. Changes in the blood during amphibian metamorphosis, pp. 461-490. In: L.I. Gilbert, and E. Frieden [eds.]. Metamorphosis: A Problem in Developmental Biology, 2nd ed. Plenum Press, NY, USA.

Budgett, J.S. 1900. Observations on *Polypterus* and *Protopterus. Proc. Cambr. Philos. Soc.* 10: 236-240.

Burggren, W.W. 1989. Lung structure and function: amphibians, pp. 153-192. In: S.C. Wood [ed.]. Comparative Pulmonary Physiology: Current Concepts. Marcel Dekker, NY, USA.

Burggren, W.W., and A.Mwalukoma. 1983. Respiration during chronic hypoxia and hyperoxia in larval and adult bullfrogs (*Rana catesbeiana*). I. Morphological responses of lungs, skin and gills. *J. Exper. Biol.* 105: 191-203.

Burggren, W.W., M.E. Feder, and A.W. Pinder. 1983. Temperature and the balance between aerial and aquatic respiration in larvae of *Rana berlandieri* and *Rana catesbeiana. Physiol. Zool.* 56: 263-273.

Claussen, C.P., and A. Hue. 1987. Light and electronmicroscopic studies of the lung of *Triturus alpestris* (Laurenti) (Amphibia). *Zool. Anz.* 218: 115-128.

Collazo, A. 1993. Evolutionary correlations between early developmant and life history in plethodontid salamanders and teleost fishes. *Amer. Zool.* 33: 60A.

Czopek, J. 1962b. Smooth muscle in the lungs of some urodeles. *Nature, Lond.* 193: 798-811.

D'Aoust, G. 1970. The role of lactic acid in gas secretion in the teleost swim bladder. *Comp. Biochem. Physiol.* 32: 637-668.

De Jong, H.J., and C Gans. 1969. On the mechanism of respiration in the bullfrog, *Rana catesbeiana:* reassessment. *J. Morph.* 127: 259-290.

De Saint-Aubain, M.L., and K. Wingstrand. 1981. A sphincter in the pulmonary artery of the frog *Rana temporaria* and its influence on blood flow in skin and lungs. *Acta. Zool. Stockh.* 60: 163-172.

Driedzic, W.R., and H. Gesser. 1994. Energy metabolism and contractility in ectothermic vertebrate hearts: hypoxia, acidosis, and low temperature. *Physiol. Rev.* 74: 221-258.

Duellman, W.E, and L. Trueb. 1986. Biology of Amphibians. McGraw-Hill Book Com., NY, USA.

Emilio, M.G., and G. Shelton. 1974. Gas exchange and its effect on blood-gas concentrations in the amphibian, *Xenopus laevis. J. Exper. Biol.* 60: 567-579.

Foxon, G.E.H. 1964. Blood and respiration, pp. 151-209. In: J.A. Moore [ed.]. Physiology of Amphibia. Academic Press, NY, USA.

Gans, C. 1970. Respiration in the early tetrapods: the frog is a red herring. *Evolution* 24: 740-751.

Gans, C. 1971. Strategy and sequence in the evolution of the external gas exchangers of ectothermal vertebrates. *Forma Functio* 3: 66-104 .

Gans, C. 1976. Ventilatory mechanisms and problems in some amphibious aspiration breathers (*Chelydra*, Caiman- Reptilia), pp. 357-374. In: G.M. Hughes [ed.]. Respiration of Amphibious Vertebrates. Academic Press, London.

Gatzy, J.T. 1975. Ion transport across the excised bullfrog lung. *Amer. J. Physiol.* 228: 1162-1171.

Glass, M.L., W.W. Burggren, and K. Johansen. 1981. Pulmonary diffusing capacity of the bullfrog (*Rana catesbeiana). Acta. Physiol. Scand.* 113: 485-490.

Goldie, R.G., J.F. Bertram, A. Warton, J.M. Papadimitriou, and J.W. Paterson. 1983. Pharmacological and ultrastructural study of alveolar contractile tissue in toad (*Bufo marinus*) lung. *Comp. Biochem. Physiol.* 75: 343-349.

Goniakowska, L. 1973. Metabolism, resistance to hypotonic solutions and ultrastructure of erythrocytes of five amphibian species. *Acta. Biol., Cracov.* 16: 114-123.

Goniakowska-Witalinska, L. 1974. Respiration, resistance to hypotonic solutions and ultrastructure of erythrocytes of *Salamandra salamandra. Bull. Acad. Polon. Sci.* 22: 59-75.

Goniakowska-Witalinska, L. 1978. Ultrastructure and morphometric study of the lung of the European salamander, *Salamandra salamandra* L *Cell Tissue Res.* 191: 343-356.

Goniakowska-Witalinska, L. 1980. Ultrastructural and morphometric changes in the lung of newt, *Triturus cistatus carnifex* (Laur.) during ontogeny. *J. Anat.* 130: 571-583.

Goniakowska-Witalinska, L. 1980. Scanning and transmission electron microscopic study of the lung of the newt, *Triturus alpestris* Laur. *Cell Tissue Res.* 205: 133-145.

Goniakowska-Witalinska, L. 1982. Develoment of the larval lung of *Salamandra salamandra* L *Anat. Embryol.* 164: 113-137.

Goniakowska-Witalinska, L. 1986. Lung of the tree frog, *Hyla arborea:* A scanning and transmission electron microscope study. *Anat. Embryol.* 174: 379-389.

Goniakowska-Witalinska, L. 1995. The histology and ultrastructure of the amphibian lung, pp. 77-112. In: L.M. Pastor [ed.]. Histology, Ultrastructure and Immunohistochemistry of the Respiratory Organs in non-mammalian Vertebrates. Publicaciones Universidad de Murcia, Murcia, Spain.

Grigg, G.C. 1965. Studies on the Queensland lungfish, *Neoceratodus forsteri* (Krefft). *Aust. J. Zool.* 13: 243-257.

Hakim, A., J.S.D. Munshi, and G.M. Hughes. 1978. Morphometrics of the respiratory organs of the Indian green snakehead fish, *Channa punctata. J. Zool., Lond.* 184: 519-543.

Hutchison, V.H. 1968. Relation of body size and surface area to gas exchange in anurans. *Physiol. Zool.* 41: 65-85.

Jackson, D.C. 1987. How do amphibians breathe both water and air, pp. 49-58. In: P. Dejours, L. Bolis, C.R. Taylor, and E.R. Weibel [eds.]. Comparative Physiology: Life in Water and on Land. Springer-Verlag, Berlin.

Kruhoffer, M.L., A.S. Abe, and K. Johansen. 1987. Control of breathing in an amphibian *Bufo paracnemius:* effects of temperature and hypoxia. *Respir. Physiol.* 69: 267-275.

Maina, J.N. 1998. The Gas Exchangers: Structure, Function and Evolution of the Respiratory Processes. Springer-Verlag, Heidelberg.

Maina, J.N. 1989. The morphology of the lung of the East African tree frog *Chiromantis petersi* with observations on the skin and the buccal cavity as secondary gas exchange organs: a TEM and SEM study. *J. Anat.* 165: 29-43.

Maina, J.N., and G.M.O. Maloiy. 1988. A scanning and transmission electron microscopic study of the lung of a caecilian *Boulengerula taitanus. J. Zool., Lond.* 215: 739-751.

Martin, K.M., and V.H. Hutchison. 1979. Ventilatory activity in *Amphiuma tridactylum* and *Siren lacertina* (Amphibia, Caudata). *J. Herpatol.* 13: 427-434.

McClanahan, L.L., R. Rodolfo, and V.H. Shoemaker. 1994. Frogs and toads in deserts. *Sci. Amer.* 273: 82-88.

Meban, C. 1977. Ultrastructure of the respiratory epithelium in the lungs of the newt *Triturus cristatus. Acta. Zool. Stockh.* 58: 151-167.

Meban, C. 1979. An electron microscopy study of the respiratory epithelium in the lungs of the fire salamander (*Salamandra salamandra*). *J. Anat.* 128: 215-221.

Meban, C. 1980. Thicknesses of the air-blood barriers in vertebrate lungs. *J. Anat.* 131: 299-307.

Milner, A.R. 1988. The relationships and origins of living amphibians, pp. 59-102. In: M.J. Benton [ed.]. The Phylogeny and Classification of Tetrapods, vol 1, Smphibians, Reptiles and Birds. Clarendon Press, Oxford.

Milsom, W.K. 1989. Mechanisms of ventilation in lower vertebrates: adaptations to respiratory and non-respiratory constraints. *Can. J. Zool.* 67: 2943-2963.

Nieden, F. 1913. Gymnophiona (Amphibia: Apoda). *Tierreich* 37: 1-31.

Okada, Y., S. Ishiko, S. Daido, J. Kim, and S. Ikeda. 1962. Comparative morphology of the lung with special reference to the alveolar epithelial cells. I. Lungs of Amphibia. *Acta. Tuberc. Jap.* 11: 63-87.

Pattle, R.E. 1976. The lung surfactant in the evolutionary tree, pp. 233-255. In: G.M. Hughes [ed.]. Respiration of Amphibious Vertebrates. Academic Press, London.

Pattle, R.E., C. Schock, J.M. Creasey, and G.M. Hughes. 1977. Surpellic films, lung surfactant, and their cellular origin in newt, caecilian and frog. *J. Zool., Lond.* 182: 125-136.

Pohunkova, H., and G.M. Hughes. 1985a. Structure of the lung of the clawed toad (*Xenopus laevis* Daudin). *Folia Morphologica* XXXIII: 385-390.

Shannon, P., and D.L. Kramer. 1988. Water depth alters respiratory behaviour of *Xenopus laevis. J. Exper. Biol.* 137: 597-602.

Shelton, G. 1970. The effect of lung ventilation on blood flow to the lungs and body of the amphibian, *Xenopus laevis. Respir. Physiol.* 9: 183-196.

Shield, J.W., and P.J. Bentley. 1973. Respiration of some urodele and anuran Amphibia in water. I. Role of the skin and the gills. *Comp. Biochem. Physiol.* 46A: 17-28.

Shield, J.W., P.J. Bentley. 1973. Respiration of some urodele and anuran Amphibia in air. II. Role of the skin and lungs. *Comp. Biochem. Physiol.* 46A: 29-38.

Smit, A.W., and J.I. Flanagin. 1994. Bimodal respiration in aquatic and terrestrial apodan amphibians. *Amer. Zool.* 34: 247-263.

Smith, D.G., and Campbell. 1976. The anatomy of the pulmonary vascular bed in the toad, *Bufo marinus* and *Xenopus laevis. Cell Tissue Res.* 178: 1-14.

Smith, D.G., and L. Rapson. 1977. Differences in pulmonary microvascular anatomy between *Bufo marinus* and *Xenopus laevis. Cell Tissue Res.* 178: 1-15.

Stark-Vancs, V., P.B. Bell, and V.H. Hutchison. 1984. Morphological and pharmacological basis for pulmonary ventilation in *Amphiuma tridactylum:* an ultrastructural study. *Cell Tissue Res.* 238: 1-12.

Stinner, J.N., V.H. Shoemaker. 1987. Cutaneous gas exchange and low evaporative water loss in the frogs *Phyllomedusa sauvagei* and *Chiromantis xeraphelina. J. Comp. Physiol.* 157B: 423-427.

Taylor, E.H. 1968. The Caecilians of the World. University of Kansas Press, Lawrence, KS, USA.

Tenney, S.M., and J.B. Tenney. 1970. Quantitative morphology of cold blooded lungs: Amphibia and Reptilia. *Respir. Physiol.* 9: 197-215.

Tilley, S.G., and J. Bernado. 1993. Life history evolution in plethodontid salamanders. *Herpetologia* 49: 154-163.

Toews, D.P., and D. MacIntyre. 1978. Respiration and circulation in an apodan amphibian. *Can. J. Zool.* 56: 199-214.

Toews, D.P., and R.G. Boutilier. 1986. Acid-base regulation in the amphibia, pp. 266-308. In: N. Heisler [ed.]. Acid-base Regulation in Animals. Amsterdam, Elsevier.

Vergara, A.G., and G.M. Hughes. 1981. Phospholipids in washings from the lung of the frog (*Rana pipiens*). *J. Comp. Physiol.* 27: 117-120.

Wake, M.H. 1974. The comparative morphology of the caecilian lung. *Anat. Rec.* 178: 483.

Wake, M.H. 1977. The reproductive biology of caecilians: an evolutionary perspective, pp. 73-101. In: D.H. Taylor, and S.I. Guttman [eds.]. The Reproductive Biology of Amphibians. Plenum Press, NY, USA.

Wake, M.H., and G. Roth. 1989. Paedomorphosis: new evidence for its importance in salamander evolution. *Amer. Zool.* 29: 134A.

Wake, M.H., S.B. Marks. 1993. Development and evolution of plethodontid salamanders: a review of prior studies and a prospectus for future research. *Herpetologia* 49: 194-203.

Wasserzug, R.J., R.D. Paul, and M.E. Feder. 1981. Cardiorespiratory synchrony in anuran larvae (*Xenopus laevis, Pachymedusa dacnicolor,* and *Rana berlandieri*). *Comp. Biochem. Physiol.* 70A: 329-334.

Welsch, U. 1983. Phagocytosis in the amphibian lung. *Anat. Anz.* 154: 323-327.

West, N.H., and D.R. Jones 1975. Breathing movements in the frog *Rana pipiens*. I. The mechanical events associated with lung and buccal ventilation. *Can. J. Zool.* 53: 332-344.

West, N.H., and W.W. Burggren. 1983. Reflex interactions between aerial and aquatic gas exchange organs in larval bullfrogs. *Amer. J. Physiol.* 244: R770-R777.

West, N.H., and W.W. Burggren. 1984. Control of pulmonary and cutaneous blood flow in the toad, *Bufo marinus*. *Amer. J. Physiol.* 247: R884-R894.

Whitford, R.W., and V.H. Hutchison. 1967. Body size and metabolic rate in salamanders. *Physiol. Zool.* 40: 127-133.

Withers, P.C., S.S. Hillman, L.A. Simmons, and A.C. Zygmut. 1988. Cardiovascular adjustments to enforced activity in the anuran amphibian, *Bufo marinus*. *Comp. Biochem. Physiol.* A89: 45-49.

Wood, S.C., R.E. Weber, G.M.O. Maloiy, and K. Johansen. 1975. Oxygen uptake and blood respiratory properties of the caecilian *Boulengerula taitanus*. *Respir. Physiol.* 24: 355-363.

7   REPTILIAN LUNG

Becker, H.O., W. Bohme, and S.F. Perry. 1989. The lung morphology of lizards (Reptilia:Varaniidae) and its taxonomic-phylogenetic meaning. *Bonn. Zool. Beitr.* 40: 27-56.

Behrensmeyer, A.K., and S.M. Kidwell. 1985. Taphonomy and paleobiology. *Paleobiology* 11: 105-119.

Bennett, A.F. 1982. The energetics of reptilian activity, pp. 155-199. In: C. Gans, and F.H. Pough [ed.]. Biology of the Reptilia, Physiological Ecology. Academic Press, NY, USA.

Benton, M.J. 1993. Late Triassic extinctions and the origin of the dinosaurs. *Science* 260: 769-770.

Bozinovic, F. 1993. Scaling basal and maximal metabolic rate in rodents and the aerobic capacity model for the evolution of endothermy. *Physiol. Zool.* 65: 921-932.

Carroll, R.L. 1970. Quantitative aspects of the amphibian-reptilian transition. *Forma Functio* 3: 165-178.

Cope, E.D. 1894. On the lungs of Ophidia. *Amer. Philos. Soc.* 33: 217-224.

Denison, R.H. 1941. The soft anatomy of *Bothriolepis*. *J. Paleontol.* 15: 553-561.

Else, P.L., and A.J. Hubert. 1981. Comparison of the 'mammal machine' and the 'reptilian machine' energy production. *Amer. J. Physiol.* 240: R3-R9.

Else, P.L., and A. J. Hubert. 1983. A comparative study of the metabolic capacity of hearts from reptiles and mammals. *Comp. Biochem. Physiol.* 76A: 553-557.

Gans, C, G.M. Hughes. 1967. The mechanism of lung ventilation in the tortoise *Testudo graeca* Linne. *J. Exper. Biol.* 47: 1-20.

Gans, C., and B. Clark. 1978. Air flow in reptilian ventilation. *Comp. Biochem. Physiol.* 60A: 453-457.

Gatz, R.N., M.L. Glass, and S.C. Wood. 1987. Pulmonary function of the green sea turtle, *Chelonia mydas*. *J. Appl. Physiol.* 62: 459-463 .

Gaunt, A.S., and C. Gans. 1969. Mechanics of respiration in the snapping turtle, *Chelydra serpentina* (Linne). *J. Morph.* 128: 195-228.

George, J.C., and R.V. Shah. 1956. Comparative morphology of the lung in snakes with remarks on the evolution of the lung in reptiles. *J. Anim. Morphol. Physiol.* 3: 1-7.

George, J.C., and R.V. Shah. 1965. Evolution of air sacs in Sauropodia. *J. Anim. Morphol. Physiol.* 12: 255-263.

Glass, M.L., and K. Johansen. 1981. Pulmonary diffusing capacity in reptiles (relations to temperature and oxygen uptake). *J. Comp. Physiol.* 107: 169-178.

Glass, M.L., and S.C. Wood. 1983. Gas exchange and control of breathing in reptiles. *Physiol. Rev.* 63: 232-260.

Glass, M.L., S.C. Wood, R.W. Hoyt, and K. Johansen. 1979. Chemical control of breathing in the lizard *Varanus exanthematicus*. *Comp. Biochem. Physiol.* 62A: 999-1003.

Gratz, R.K., A. Ar, J. Geiser. 1981. Gas tension profile of the lung of the viper, *Vipera xanthina palestinae*. *Respir. Physiol.* 44: 165-171.

Hlastala, M.P., T.A. Standaert, D.J. Pierson, and D.L. Luchtel. 1985. The matching of ventilation and perfusion in the lung of the tegu, *Tupinambis nigropunctus*. *Respir. Physiol.* 60: 277-294.

Klaver, C.J.J. 1973. Lung anatomy: aid in chameleon taxonomy. *Beaufortia* 20: 155-177.

Klaver, C.J.J. 1981. Lung morphology in the Chamaeleonidae (Sauria) and its bearing upon phylogeny, systematics and zoogeography. *Z. Zool. Syst. Evolutionforsch* 19: 36-58.

Klemm, R.D., R.N. Gatz, J.A. Westfall, and M.R. Fedde. 1979. Microanatomy of the lung parenchyma of a tegu lizard *Tupinambis nigropunctatus*. *J. Morph.* 161: 257-280.

Luchtel, D.L., and K.V. Kardong. 1981. Ultrastructure of the lung of the rattlesnake, *Crotalus viridis oreganus*. *J. Morph.* 169: 29-47.

Maina, J.N. 1989. The morphology of the lung of the black mamba *Dendroaspis polylepis* (Reptilia: Ophidia: Elapidae): a scanning and transmission electron microscopic study. *J. Anat.* 167: 31-46.

Maina, J.N. 1998. The Gas Exchangers: Structure, Function and Evolution of the Respiratory Processes. Springer-Verlag, Heidelberg.

Maina, J.N., C.J. Veltcamp, and J. Henry. 1999. Study of the spatial organization of the gas exchange components of a snake lung–sand boa *Eryx colubrinus* (Reptilia: Ophidia: Colubridae)–by latex casting. *J. Zool. (Lond.)* 247: 81-90.

Maina, J.N., G.M.O. Maloiy, C.N. Warui, E.K. Njogu, and E.D. Kokwaro.1989. A scanning electron microscope study of the reptilian lungs: The savanna monitor lizard (*Varanus exanthematicus*) and the pancake tortoise (*Malacochersus tornieri*). *Anat. Rec.* 224: 514-522.

Meban, C. 1978. Functional anatomy of the lungs of the green turtle, *Lacerta viridis*. *J. Anat.* 125: 421-436.

Mertens, R. 1960. The World of Amphibians and Reptiles. McGraw-Hill, NY, USA.

Milsom, W.K., and K. Johansen. 1975. The effect of buoyancy induced lung volume changes on respiratory frequency in a chelonian (*Caretta caretta*). *J. Comp. Physiol.* 98:157-160.

Pastor, L.M. 1995. The histology of the reptilian lungs, pp. 131-153. In: L.M. Pastor [ed.]. Histology, Ultrastructure and Immunohistochemistry of Respiratory Organs in Non-mammalian Vertebrates. Secretariado de Oublicaciones de la Universidad de Murcia, Spain.

Pastor, L.M., J. Ballesta, M.T. Castells, R. Perez-Tomas, J.A. Marin, and J.F. Madrid. 1989. A microscopic study of the lung of *Testudo graeca*. *J. Anat.* 162: 19-33.

Perry, S.F. 1983. Reptilian lungs: functional anatomy and evolution. *Adv. Anat. Embryol. Cell Biol.* 79: 1-81.

Perry, S.F. 1992. Gas exchange strategies in reptiles and the origin of the avian lung, pp. 149-167. In: S.C. Wood, R.E. Weber, A.R. Hargens, and R.W. Millard [eds.]. Physiological Adaptations in Vertebrates: Respiration, Circulation, and Metabolism. Marcel Dekker Inc., NY, USA.

Perry, S.F. 1999. Lungs: comparative anatomy, functional morphology, and evolution, pp. 1-92. In: C. Gans, and A.S. Gaunt [eds.]. Biology of Reptilia, vol 19. Morphology Society for the Study of Amphibians and Reptiles, Ithaca (NY).

Perry, S.F., and H.R. Duncker. 1978. Lung architecture, volume and static mechanics in five species of lizards. *Respir. Physiol.* 34: 61-81.

Phleger, C.F., D.G. Smith, D.H. Macintyre, and B.S. Saunders. 1978. Alveolar and saccular lung phospholipids of the anaconda, *Eunectes murinus*. *Can. J. Zool.* 56: 1009-1013.

Pohunkova, H., and G.M. Hughes. 1985b. Ultrastructure of the lungs of the garter snake. *Folia Morph, Prague* 23: 254-258.

Schmalhausen, I.I. 1968. The Origin of Terrestrial Vertebrates. Academic Press, London.

Seymour, R.S., R.G. Spragg, and M.T. Hartman. 1981. Distribution of ventilation and perfusion in the sea snake, *Pelamis platurus*. *J. Comp. Physiol.* 60A: 145:109-115.

Solomon, S.E., and M. Purton. 1984. The respiratory epithelium of the lung in the green turtle (*Chelonia mydas* L). *J. Anat.* 139: 353-361.

Stinner, J.N. 1982. Functional anatomy of the lung of the snake, *Pituophis melanoleucus. Amer. J. Physiol.* 243: R251-257.

Stinner, J.N. 1987. Gas exchange and air flow in the lung of the snake, *Pituophis melanoleucus. J. Comp. Physiol.* 157: 307-314.

Tenney, S.M. 1979. A synopsis of breathing mechanisms, pp. 51-106. In: S. C. Wood, and C. Lenfant [eds.]. Evolution of Respiratory Processes: A Comparative Approach. Marcel Dekker Inc., NY, USA.

Tenney, S.M., and J.B. Tenney. 1970. Quantitative morphology of cold blooded lungs: Amphibia and Reptilia. *Respir. Physiol.* 9: 197-215.

Tenney, S.M., D. Bartlett, J.P. Farber, and J.E. Remmers. 1984. Mechanics of the respiratory cycle in the green turtle (*Chelonia mydas*). *Respir. Physiol.* 22: 361-368.

Todd, G.T. 1980. Evolution of the lung and the origin of bony fishes—a casual relationship? *Amer. Zool.* 20: 757A.

Ultsch, G.R., and D.C. Jackson. 1982. Longterm submergence at 3°C of the turtle, *Chrysemys picta belii*, in normoxic and severely hypoxic water. I. Survival, gas exchange and acid-base status. *J. Exper. Biol.* 96: 11-28.

Verde, M.R. 1951. The morphology and histology of the lung in snakes. *J. Univ. Bombay* 19: 79-89.

West, N.H., P.J. Butler, and R.M. Bevan. 1992. Pulmonary blood flow at rest and during swimming in the green turtle, *Chelonia mydas. Physiol. Zool.* 65: 287-310.

White, F.N., and P.E. Bickler. 1987. Cardiopulmonary gas exchange in the turtle: a model analysis. *Amer. Zool.* 27: 31-40.

Wood, S.C., and K. Johansen. 1974. Respiratory adaptations to diving in the Nile monitor lizard, *Varanus niloticus. J. Comp. Physiol.* 89: 145-158.

Wood, S.C., K. Johansen, and R.N. Gatz. 1978. Pulmonary blood flow, ventilation-perfusion ratio, and oxygen transport in a varanid lizard. *Amer. J. Physiol.* 233: R89-93.

8 AVIAN LUNG

Bartholomew, G.A., and J.R.B. Lighton. 1986. Oxygen consumption during hover-feeding in free-ranging Anna hummingbirds. *J. Exper. Biol.* 123: 191-199.

Berger, M., and J.S. Hart. 1974. Physiology and energetics of flight, pp. 415-477. In: D.S. Farner, andd J.R. King [eds.]. Avian Biology, vol 4. Academic Press, NY, USA.

Berger, M., O.Z. Roy, and J.S. Hart. 1970. The coordination between respiration and wing beats in birds. *Z. Vergl. Physiol.* 66: 190-200.

Berker, L.V., and L.C. Marshall. 1965. The origin and rise of oxygen concentration in the Earth's atmosphere. *J. Atmos. Sci.* 22: 225-261.

Bernstein, M.H. 1990. Avian respiration and high altitude tolerance, pp. 30-40. In: J.R. Sutton, G.C. Coates, and J.E. Remmers [eds.]. Hypoxia: The Adaptations. B.C. Decker Inc., Burlington, Ontario.

Blair, H.S. 1994. Molecular evidence for the origin of birds. *Proc. Natl. Acad. Sci. USA* 91: 2621-2624.

Brakenbury, J.H. 1984. Physiological responses of birds to flight and running. *Biol. Rev.* 59: 559-575.

Brakenbury, J.H. 1987. Ventilation of the lung-air sac system, pp. 39-69. In: T.J. Seller [ed.]. Bird Respiration vol I. CRC Press, Boca Raton, FL, USA.

Brakenbury, J.H. 1991. Ventilation, gas exchange and oxygen delivery in flying and flightless bird, pp. 125-147. In: A.J. Woakes, M.K. Grieshaber, and C.R. Bridges [eds.]. Physiological Strategies for Gas Exchange and Metabolism. Cambridge University Press, Cambridge.

Breeze, R.G., and E.B. Wheeldon. 1977. The cells of the pulmonary airways. *Amer. Rev. Respir. Dis.* 116: 705-777.

Broyles, R.H. 1981. Changes in the blood during amphibian metamorphosis, pp. 461-490. In: L.I. Gilbert, and E. Frieden [eds.]. Metamorphosis: A Problem in Developmental Biology, 2nd ed. Plenum Press, NY, USA.

Butler, P.J. 1991. Exercise in birds. *J. Exper. Biol.* 160: 233-262.

Butler, J., and A.J. Woakes. 1980. Heart rate, respiratory frequency and wing beat frequency of free flying barnacle geese, *Branta leucopsis. J. Exper. Biol.* 85: 213-226.

Cracraft, J.A. 1986. The origin and early diversification of birds. *Paleobiology* 12: 383-399.

De Beers, G. 1954. *Archeopteryx lithographica.* Br. Mus. Nat Hist. London.

Dubach, M. 1981. Quantitative analysis of the respiratory system of the house sparrow, budgerigar, and violet-eared hummingbird. *Respir. Physiol.* 46: 43-60.

Duncker, H.-R. 1972. Structure of avian lungs. *Respir. Physiol.* 14: 44-63.

Duncker, H.-R. 1974. Structure of the avian respiratory tract. *Respir. Physiol.* 22: 1-34.

Duncker H.-R. 1978. Development of the avian respiratory system, pp. 260-273. In: J. Piiper [ed.]. Respiration in Birds, Adult and Embryonic. Springer-Verlag, Berlin.

Duncker, H.R., and M. Guntert. 1985. The quantitative design of the avian respiratory system: from hummingbird to the mute swan, pp. 361-378. In: W. Nachtigall [ed.]. BIONA Report No. 3. Gustav-Fischer Verlag, Stuttgart.

Fedde, M.R. 1980. The structure and gas-flow pattern in the avian respiratory system. *Poult. Sci.* 59: 2642-2653.

Feducia, A. 1980. The Age of Birds. Harvard University Press, Cambridge, MA, USA.

Jones, J.H., E.L. Effmann, and K. Schmidt-Nielsen 1985. Lung volume changes during respiration in ducks. *Respir. Physiol.* 59: 15-25.

King, A.S. 1966. Structural and functional aspects of the avian lung and its air sacs. *Int Rev. Gen. Exp. Zool.* 2: 171-267.

Laybourne, R.C. 1974. Collision between a vulture and an aircraft at an altitude of 37,000 ft. *Wilson Bull.* 86: 461-462.

Macklem, P.T., P. Bouverot, and P. Scheid. 1979. Measurement of the distensibility of the parabronchi in duck lungs. *Respir. Physiol.* 38: 23-35.

Maina, J.N. 1982. A scanning electron microscopic study of the air and blood capillaries of the lung of the domestic fowl (*Gallus domesticus*). *Experientia* 35: 614-616.

Maina, J.N. 1984. Morphometrics of the avian lung. 3. The structural design of the passerine lung. *Respir. Physiol.* 55: 291-309.

Maina, J.N. 1988. Scanning electron microscopic study of the spatial organization of the air- and blood conducting components of the avian lung. *Anat. Rec.* 222: 145-153.

Maina, J.N. 1989. Morphometrics of the avian lung, pp. 307-368. In: A. S. King, and J. McLelland [eds.] Form and Function in Birds, vol. 4. Academic Press, London.

Maina, J.N. 1993. Morphometrics of the avian lung: the structural-functional correlations in the design of the lungs of birds. *Comp. Biochem. Physiol.* 105: 397-410.

Maina, J.N. 1996. Perspectives on the structure and function in birds, pp. 163-256. In: E. Rosskoff [ed.]. Diseases of Cage and Aviary Birds. Williams and Wilkins, Baltimore, MD, USA.

Maina, J.N. 1998. The Gas Exchangers: Structure, Function and Evolution of the Respiratory Processes. Springer-Verlag, Heidelberg.

Maina, J.N. 2000. What it takes to fly: the novel respiratory structural and functional adaptations in birds and bats. *J. Exper. Biol.* 203: 3045-3064.

Maina, J.N., and A.S. King. 1982. The thickness of the avian blood-gas barrier: qualitative and quantitative observations. *J. Anat.* 134: 553-562.

Maina, J.N., A.S. King, G. Settle. 1989. An allometric study of the pulmonary morphometric parameters in birds, with mammalian comparison. *Phil. Trans. R. Soc. Lond.* 326B: 1-57.

Norberg U.M. 1990. Vertebrate Flight: Mechanics, Physiology, Morphology, Ecology and Evolution. Springer-Verlag, Berlin–Heidelberg.

Olmo, E. 1991. Genome variations in the transition from amphibians to reptiles. *J. Mol. Evol.* 33: 68-75.

Perry, S.F. 1992. Gas exchange strategies in reptiles and the origin of the avian lung, pp. 149-167. In: S.C. Wood, R.E. Weber, A.R. Hargens, and R.W. Millard [eds.]. Physiological Adaptations in Vertebrates: Respiration, Circulation, and Metabolism. Marcel Dekker Inc., NY, USA.

Piiper, J. [ed.]. 1978. Respiratory Function in Birds, Adult and Embryonic. Springer, Berlin-Heidelberg.

Powell, F. L., and P. Scheid. 1989. Physiology of gas exchange in the avian respiratory system, pp. 393-437. In: A.S. King, and J. McLelland [eds.]. Form and Function in Birds, vol 4. Academic Press, London.

Salomonsen, F. 1967. Migratory movements of the Arctic tern (*Sterna paradisea pontoppidan*) in the Southern Ocean. *Det. Kgl. Danske Vid Selsk, Biol. Med.* 24: 1-37.

Scheid, P. 1979. Mechanisms of gas exchange in bird lungs. *Rev. Physiol. Biochem. Pharmacol.* 86: 137-186.

Scheid, P. 1990. Avian respiratory system and gas exchange, pp. 4-7. In: J. R. Sutton, G. Coates, and J. E. Remmers [eds.]. Hypoxia: The Adaptations. BC Decker Inc., Burlington, Ontario.

Scheid, P., and J. Piiper. 1972. Cross-currrent gas exchange in the avian lungs: effects of reversed parabronchial air flow in ducks. *Respir. Physiol.* 16: 304-312.

Scheid, P., and J. Piiper. 1989. Respiratory mechanics and air flow in birds, pp. 369-391. In: A.S. King, and J. McLelland [ed.]. Form and Function in Birds, vol 4. Academic Press, London.

Suarez, R.K. 1992. Hummingbird flight: sustaining the highest mass-specific metabolic rates among vertebrates. *Experientia* 48: 565-570.

Torre-Bueno, J.R. 1985. The energetics of avian flight at altitude, pp. 45-87. In: W. Nachtigall [ed.]. Bird flight, BIONA Report 3. Gustav Fischer, Stuttgart.

Tucker, V.A. 1974. Energetics of natural avian flight, pp. 298-333. In: R.A. Paynter [ed.]. Avian Energetics. Nuttal Ornithological Club, Cambridge, MA, USA.

Wells, D.J. 1993. Ecological correlates of hovering flight of hummingbirds. *J. Exper. Biol.* 178: 59-70.

9 MAMMALIAN LUNG

Banzett, R.B., C.S. Nations, N. Wang, P.J. Butler, and J.L. Lehr. 1992. Mechanical interdependence of wing beat and breathing in starlings. *Respir. Physiol.* 89: 27-36.

Bartholomew, G.A., P. Leitner, and J.E. Nelson. 1964. Body temperature, oxygen consumption, and heart rate in three species of Australian flying foxes. *Physiol. Zool.* 37: 179-198.

Beitinger, T.L., and M.J. Pettit. 1984. Comparison of low oxygen avoidance in a bimodal breather, *Erpetoichthys calabaricus* and an obligate water breather, *Percina caprodes. Environ. Biol. Fishes.* 11: 235-240.

Berger, M., O.Z. Roy. and J.S. Hart. 1970. The coordination between respiration and wing beats in birds. *Z. Vergl. Physiol.* 66: 190-200.

Brain, J.D. 1985. Macrophages in the respiratory tract, pp. 447-471. In: A.P. Fishman, and A.B. Fisher [eds.]. Handbook of Physiology, vol. 1. Circulation and Nonrespiratory Functions. American Physiological Society, Bethesda, Maryland.

Burri, P.H. 1984. Lung development and histogenesis, pp. 1-46. In: A.P. Fishman, and A.B. Fischer [eds.]. Handbook of Physiology; Respiration, vol. 4. American Physiological Society Press, NY, USA.

Burri, P.H. 1985. Morphology and respiratory function of the alveolar unit. *Int. Arch. Allergy Appl. Immunol.* 76: 2-12.

Carpenter, R.E. 1986. Flight physiology of intermediate sized fruit-bats (Family: Pteropodidae). *J. Exper. Biol.* 120: 79-103.

Comroe, J.H. 1974. Physiology of Respiration: An Introductory Text. Year Book Medical Publishers, Chicago, ILL, USA.

Constantinopol, M., J.H. Jones, E.R. Weibel, H. Hoppelar, A. Lidholm, and R.H. Karas. 1989. Oxygen transport during exercise in large mammals. II. Oxygen uptake by the pulmonary gas exchanger. *J. Appl. Physiol.* 67: 871-878.

Fenton, M.B., R.M. Bringham, A.M. Mills, and I.L. Rautenbach. 1985. The roosting and foraging areas of *Epomophorus wahlbergi* (Pteropodida) and *Scotophilus viridis* (Vespertilionide) in Kruger National Park, South Africa. *J. Mammal.* 66: 461-468.

Gehr, P., M. Bachofen, and E.R. Weibel. 1978. The normal human lung: ultrastructure and morphometric estimation of diffusion capacity. *Respir. Physiol.* 32: 121-140.

Gehr, P., S. Sehovic, P.H. Burri, H. Claasen, and E.R. Weibel. 1980. The lung of shrews: morphometric estimation of diffusion capacity. *Respir. Physiol.* 44: 61-86.

Gehr, P., D.K. Mwangi, A. Amman, G.M.O. Maloiy, C.R. Taylor, and E.R. Weibel. 1981. Design of the mammalian respiratory system: V. Scaling morphometric diffusing capacity to body mass: wild and domestic animals. *Respir. Physiol.* 44: 61-86.

Glazier, J.B., J. M.B. Hughes, J.E. Maloney, and J.B. West. 1967. Vertical gradient of alveolar size in lungs of dogs frozen intact. *J. Appl. Physiol.* 23: 694-705.

Golde, L.M. G., J.J. Batenburg, and B. Robertson. 1994. The pulmonary surfactant system. *News Physiol. Sci.* 9: 13-20.

Heinemann, H.O., and A.P. Fishman. 1969. Nonrespiratory functions of mammalian lung. *Physiol. Rev.* 49: 1-61.

Hills, B.A. 1988. The Biology of the Surfactant. Cambridge University Press, Cambridge.

Jürgens, J.D., H. Bartels, and R. Bartels. 1981. Blood oxygen transport and organ weight of small bats and small nonflying mammals. *Respir. Physiol.* 45: 243-260.

Karas, R.H., C.R. Taylor, J.H. Jones, S.L. Linstedt, R.B. Reeves, and E.R. Weibel. 1987a. Adaptive variation in the mammalian respiratory system in relation to energetic demand. VII. Flow of oxygen across the pulmonary gas exchanger. *Respir. Physiol.* 69: 101-115.

King, A.S., and J. McLelland. [eds.]. 1989. Form and Function in Birds, vol 4. Academic Press, London.

Krenz, G.S., J.H. Linehan, and C.A. Dawson. 1992. A fractal continuum model of the pulmonary arterial tree. *J. Appl. Physiol.* 72: 2225-2237.

Lechner, A.J. 1984. Pulmonary design in microchiropteran lung (*Pipistrellus subflavus*) during hibernation. *Respir. Physiol.* 59: 301-312.

Maina, J.N. 1988. The morphology and morphometry of the normal lung of the adult vervet monkey *Cercopithecus aethiops*. *Amer. J. Anat.* 183: 258-267.

Maina, J.N. 1990. The morphology and morphometry of the lung of the lesser bushbaby *Galago senegalensis*. *J. Anat.* 172: 129-144.

Maina, J.N. 1998. The Gas Exchangers: Structure, Function and Evolution of the Respiratory Processes. Springer-Verlag, Heidelberg.

Maina, J.N., and A.S. King. 1984. The structural functional correlation in the design of the bat lung. A morphometric study. *J. Exper. Biol.* 111: 43-63.

Maina, J.N., S.P. Thomas, and D.M. Dallas. 1991. A morphometric study of bats of different size: correlations between structure and function of the chiropteran lung. *Phil. Trans. R. Soc. Lond.* 333B: 31-50.

Maina, J.N., G.M.O. Maloiy, and A.M. Makanya. 1992. Morphology and morphometry of the lungs of two East African mole rats, *Tachyoryctes splendens* and *Heterocephalus glaber* (Mammalia, Rodentia). *Zoomorphology* 112: 167-179.

Makanya, A.N., and J.N. Maina. 1994. Comparative morphology of the gastrointestinal tract of fruit and insect-eating bats. *Afr. J. Ecol.* 32: 158-168.

Pettigrew, J.D., B.G.M. Jamieson, S.K. Robson, L.S. Hall, K.I. McNally, and H.M. Cooper. 1987. Phylogenetic relations between microbats, megabats and primates (Mammalia: Chiroptera and Primates). *Phil. Trans. R. Soc. Lond.* 325B: 489-559.

Riedesel, M.L., and B.A. Williams. 1976. Continuous 24 hr oxygen consumption studies of *Myotis velifer.* *Comp. Biochem. Physiol.* 54A: 95-99.

Stratton, C.J. 1984. Morphology of surfactant producing cells and of the alveolar lining layer, pp. 67-118. In: B. Robertson, L.M.G. van Golde, and J.J. Batenburg [eds.]. Pulmonary Surfactant. Elsevier Science Publishers, Amsterdam.

Taylor, C.R., and E.R. Weibel. 1981. Design of the mammalian respiratory system. *Respir. Physiol.* 11: 1-10.

Tenney, S.M., and J.E. Remmers. 1963. Comparative morphology of the lung: diffusing area. *Nature, Lond.* 197: 54-56.

Thewissen, J.M.G., and S.K. Babcock. 1992. The origin of flight in bats: to go where no mammal has gone before. *BioScience* 42: 340-345.

Thomas, D.W. 1983. The annual migrations of three species of West African fruit bats (Chiroptera: Pteropodidae). *Can. J. Zool.* 61: 2266-2272.

Thomas, S.P. 1987. The physiology of bat flight, pp. 75-99. In: M.B. Fenton, P.Racey, and J.M.V. Rayner [eds.]. Recent Advances in the Study of Bats. Cambridge University Press, Cambridge.

Van Valen, L. 1979. The evolution of bats. *Evol. Theory* 4: 103-121.

Weibel, E.R. 1963. Morphometry of the Human Lung. Springer, Berlin.

Weibel, E.R. 1979. Oxygen demand and size of respiratory structures in mammals, pp. 289-346. In: S.C. Wood, and C. Lenfant [eds.]. Evolution of Respiratory Processes: A Comparative Approach. Marcel Dekker Inc., NY, USA.

Weibel, E.R. 1980. Design and structure of the human lung, pp. 224-271. In: A.P. Fishman [ed.]. Pulmonary Diseases and Disorders. McGraw-Hill, NY, USA.

Weibel, E.R. 1984. The Pathways for Oxygen. Harvard University Press, Cambridge, MA, USA.

Weibel, E.R. 1984. Lung cell bology, pp. 47-91. In: A.P. Fishman, and A.B. Fisher [eds.]. Handbook of Physiology: Respiration, vol III, sect 2. American Physiological Society, DC, USA.

Weibel, E.R. 1986. Functional morphology of lung parenchyma, pp. 89-111. In: A.P. Fishman, and A.B. Fisher [eds.]. Handbook of Physiology. The Respiratory System. Mechanics of Breathing, sect 3, vol III. American Physiological Society, Bethesda, MD, USA.

Weibel, E.R. 1991. Fractal geometry: a design principle for living organisms. *Amer. J. Physiol.* 261: L361-L369.

West, J.B. [ed.]. 1977. Bioengineering Aspects of the Lung. Marcel Dekker Inc., NY, USA.

10 SUMMARY AND CONCLUSIONS

Black, C.P., S.M. Tenney, and M.V. Kroonenburg. 1978. Oxygen transport during progressive hypoxia in bar-headed geese (*Anser indicus*) acclimated to sea level and 5600 m, pp. 79-83. In: J. Piiper [ed.]. Respiratory Function in Birds, Adult and Embryonic. Springer-Verlag, Berlin.

Faraci, M.F., and M.R. Fedde. 1986. Regional circulatory responses to hypocapnia and hypercapnia in bar-headed geese. *Amer. J. Physiol.* 250: R499-R504.

Faraci, M.F., D.L. Kilgore, and M.R. Fedde. 1984. Oxygen delivery to the heart and brain during hypoxia: Pekin duck vs bar-headed geese. *Amer. J. Physiol.* 247: R69-R75.

Fedde, M.R., J.A. Orr, H. Shams, and P. Scheid. 1989. Cardiopulmonary function in exercising bar-headed geese during normoxia and hypoxia. *Respir. Physiol.* 77: 239-262.

Geelhaar, A., and E.R. Weibel. 1971. Morphometric estimation of pulmonary diffusing capacity. III. The effect of increased oxygen consumption in Japanese waltzing mice. *Respir. Physiol.* 11: 354-366.

Jones, J.H., K.E. Longworth, A. Lindholm, K.E. Conley, and R.H. Karas RH et al. 1989. Oxygen transport during exercise in large mammals. I. Adaptive variation in oxygen demand. *J. Appl. Physiol.* 67: 862-870.

Maina, J.N. 1993. Morphometries of the avian lung: the structural-functional correlations in the design of the lungs of birds. *Comp. Biochem. Physiol.* 105: 397-410.

Maina, J.N. 1994. Comparative pulmonary morphology and morphometry: The functional design of respiratory systems, pp. 111-232. In: R. Gilles [ed.]. Advances in Comparative and Environmental Physiology, vol 20. Springer-Verlag, Heidelberg.

Maina, J.N. 1996. Perspectives on the structure and function in birds, pp. 163-256. In: E. Rosskoff [ed.]. Diseases of Cage and Aviary Birds. Williams and Wilkins, Baltimore.

Maina, J.N. 1997. The lungs of the volant vertebrates–birds and bats: How are they relatively structurally optimized for this elite mode of locomotion?, pp. 177-185. In: E.R. Weibel, C.R. Taylor, and L. Bolis [eds.]. Principles of Animal Design: The Optimization and Symmorphosis Debate. Cambridge University Press, London.

West, J.B. 1983. Climbing Mt Everest without oxygen: an analysis of maximal exercise during extreme hypoxia. *Respir. Physiol.* 52: 265-274.

Index

Accessory respiratory organ 9, 11, 21, 46, 61, 65, 69

Acid-base balance 31, 39, 40, 85, 113, 131

Acidosis 8, 31,

Acipenser transmontanus 17

Acrochomdus javanicus 45

Actinopterygian 46, 61

Adaptation 5, 9, 10, 40, 46, 60, 61, 70, 131, 133

Adenosine triphosphate 3

Adrenaline 31

Adrenergic nerves 33

Afferent vessels 20, 31, 32, 33

African 18, 20, 61, 69

African rock martin 115

Agnatha 20

Air 15

Air bladder 57

Air blood barrier 7, 18, 40, 69

Air breather 7, 61, 63

Air breathing 9, 10, 11, 15, 18, 20, 21, 31, 39, 40, 60, 61, 69, 70, 72, 85, 101, 102, 133

Air capillaries 18, 106-114, 132, 133

Air cell 62, 63, 71, 72, 75, 76, 87, 90-92, 96, 98

Air duct 86

Air flow 86, 102, 113

Air sac 11, 16, 17, 93, 101, 102, 113, 130, 132

Air way 57, 85, 107, 116, 117

Alcolapia grahami 15, 25, 26-30, 32-36, 43, 44, 47, 57, 58

Alligator 93

Alligator mississippiensis 93

Alveolar air 15

Alveolar capillaries 113

Alveolar spaces 113

Alveolar surface 10, 117

Alveoli 1, 18, 117-120, 130

Amazon 5, 20, 61

Amia calva 38, 46

Ammonia 8, 31, 33, 69, 134

Amniota 61

Amniote 8, 31, 33, 69, 134

Amphibia 2, 3, 8, 9, 11, 18-20, 39, 40, 45, 47, 60, 70, 74, 84-86, 93, 102, 114, 133

Amphipnous cuchia 69

Amphiuma means 40, 73, 74

Anabas testudineus 21, 69

Anaerobe 3

Analogy 9, 113

Anamniote 19

Ancistrus 66

Angiotensin 8, 31

Anguilla anguilla 3, 31, 57, 59

Anguilla vulgaris 40

Anoxia 8

Anser indicus 132

Anura 3, 22, 70, 72

Anus 69

Aorta 21, 47

Apoda 20, 70, 72

Aquatic 2, 9, 10, 20, 39, 40, 45, 47, 57, 60, 61, 69, 70, 73, 93, 133

Arapaima gigas 47, 60

Archeopteryx lithographica 101

Arctic tern 102

Aristotle 4, 17

Arterial blood 1, 8, 15, 33, 47, 114

Arterialization 9, 15, 114

Arteriole 4, 17

Artery 32, 33

Arthropod 10

Asphyxia 9, 10, 31

Astylosternus robustus

Atmosphere 2, 10-12, 46, 57, 60

Atria 105-107, 111, 112, 114

Avian 8, 15, 17, 18, 93, 101, 102, 113-116, 132, 133

Avian lung 15

Baboon 124-26

Bacteria 2, 3

Balaena mysticetus 18

Bar-headed goose 132

Barrier 6-8, 11, 17, 18, 20, 21, 40, 64, 69, 72, 73, 93, 102, 113, 115, 117, 130, 132

Basement membrane 8, 117

Bat 17, 18, 102, 116-118, 123, 127, 128, 130, 132, 133

Batrachophrynus 73

Bee 5

Bicarbonate 12, 57

Bimodal breather 9, 11, 22, 33, 40, 60, 61, 64-68, 133

Bionics 5

Bird 1, 3, 9, 15-18, 39, 86, 93, 101, 102, 113- 117, 130, 132, 133,

Black mamba 89, 93-99

Black-headed gull 108

Blood 1, 6-11, 13, 15-21, 31, 33, 39, 40, 45, 47, 51, 57, 61, 63, 64, 69, 72-74, 93, 113

Blood capillaries 41, 42, 45, 51, 57, 73, 74, 76, 77, 95, 96, 106-112, 113, 114, 119, 126

Blood flow 13, 15, 111

Blood vessel 20, 32-34, 45, 48-50, 61, 74, 75, 80, 98, 105, 106

Blood volume 16, 40, 113-115, 130

Blood-gas barrier 8, 11, 15, 18, 64, 72, 93, 113, 115, 117, 118, 130

Bluefish 21

Body mass 18, 70, 73, 86, 100, 113, 130, 133

Boleophthalmus boddarti 21

Bone 9, 16, 31

Bonito 21

Bothriolepis 86

Boulengerula taitanus 72-74

Bowfin 38, 61

Bradykinin 8

Branchial arch 6, 21

Breathing 1, 8, 9, 10, 15, 18-21, 39, 40, 45, 60, 61, 65, 69, 70-73, 85, 101, 102, 113, 130,

Bronchial tree 16, 17, 116

Bronchiole 118, 120

Bronchoalveolar lung 16, 116, 132

Bronchus 92, 93, 114, 118

Brush cell 93

Buccal cavity 9, 40, 43, 45, 57, 69, 72, 85, 132

Buccal force pump 57, 85

Bufo marinus 3, 20, 40, 73

Bushbaby 72, 129

Caecilian 20, 70, 71, 73, 74

Cairina moschata 3

Cambrian 2

Capacitance coefficient 10

Capillary blood 15, 114, 115, 130, 132, 133

Capillary loading 113, 114

Carassius auratus 3

Carbon dioxide 2, 5, 6, 8, 10, 12, 39, 40, 45, 47, 57, 63, 69, 85, 113

Carboniferous 2, 39, 85

Cartilage 23, 31, 74, 85

Carvia porcellus 3

Catecholamines 31

Catfish 21, 65, 69

Caudata 20, 45, 70, 73

Cell membrane 11

Cercopithecus aethiops 119, 121, 122

Chamaeleon 87

Chamaeleon chamaeleon 87

Channa punctatus 69

Channa striata 21

Chaos 131

Chiromantis petersi 41, 42, 45, 70, 71, 73, 75, 78, 81, 83

Chiroptera 117, 130

Chloride cell 21, 29, 31, 35-37

Chondrichthyan 19

Chrysemys 3

Chthonerpoton indistinctum 73

Cichlid 50

Ciliated cells 124

Circulation 8, 16, 20, 31, 61

Circulatory system 6, 7, 59, 93

Clara cell 123

Clarias batrachus 21, 23, 69

Clarias mossambicus 21, 23, 64-68, 69,

Cleidoic 85

Climbing perch 21, 69

Cobitis 66

Cocoon 64

Cod 3, 57

Coefficient 5, 10

Coelacanth 17, 60

Coelom 85, 130

Coelomic cavity 130

Coelurosaurs 86, 101

Colibri coruscans 18, 114, 115

Collagen 8, 47, 55, 56, 74, 98, 99, 117, 129

Columba livia 3

Compliance 74, 93, 100, 113

Con-current 15

Cortisol 31

Counter-current 6, 11, 13, 15, 17, 19, 57, 114, 113, 133

Cretaceous 86, 116

Crocodile 93, 100

Crocodylus niloticus 93

Cross-current 15, 17, 111, 112, 114

Crossopterygian 60, 61

Cryptobiosis 3

Cryptobranchus alleganiensis 45, 47, 72

Cuchia eel 69

Cutaneous respiration 19, 39, 40, 45

Cutaneous vasculature 40, 45

Cyanobacteria 2

Cyanocephalus volans 102

Cygnus olor 113

Danio danglia 17, 21

Dead space 11

Dendroaspis polylepis 88, 89, 93-97

Design 3, 8-10, 14-21, 31, 40, 45, 60, 84, 86, 93, 102, 114-117, 130-134

Desmosome 47

Dessication 10, 39, 133

Devonian 2, 39, 86

Diaphragm 16, 85

Dichotomous 16, 17, 116

Diffusing capacity 39, 40, 72, 86, 115, 117, 130

Diffusion 8-13, 15, 57, 61

Diffusivity 2, 10

Dinosaurs 101, 116

Dinotopterus 69

Dipnoi 20, 21, 46, 60, 61, 70

Dogfish 3

Domestic fowl 3, 101, 103-107, 109-112, 114, 115

Double capillary system 93, 114, 133

Double circulation 61

Draco volans 102

Dromiceus novaehollandiae 101, 115

Duck 3

Earth 2, 3, 19, 69, 85, 101, 117

Ectotherm 18, 24, 86, 132

Eel 3, 4, 19, 31, 40, 57, 59, 69

Egg 19, 20, 39, 70, 71, 85

Elasmobranch 19, 20, 57

Electrophorus electricus 69

Eleutherodactylus coqui 40

Emu 101, 102, 115

Endoplasmic reticulum 23, 127

Endothelial cell 21, 57, 74, 75, 78, 96, 98, 99, 117

Endotherm 1, 9, 17, 18, 86, 101, 116, 132

Endothermy 86

Energy 1-5, 10, 16-18, 57, 93, 102, 115, 130-132, 136

Entropy 131

Eocene 130

Epidermis 39, 45, 73, 85

Epithelial cell 24, 31, 47, 57, 63, 74, 93, 94, 96-98, 117, 121, 122

Epithelium 8, 17, 21, 26, 31, 33, 44, 45, 48, 93, 117

Epomophorus wahlbergi 18, 130

Eptesicus fuscus 130

Erythrocyte 13, 26, 28, 40, 45, 49, 65, 68, 75, 79, 80, 96-99, 108, 109, 127-129

Euclid 16

Eukaryotic 1

Euryhaline fish 31

Evagination 10-12, 19, 133

Evolution 1, 3, 5, 8, 9, 16, 19, 39, 46, 60, 61, 70, 85, 101, 102, 113, 116, 131, 133

Exchange tissue 1593, 105, 113, 114, 132

Exercise 1, 15, 33, 47, 73, 86, 101, 102, 114

Expiration 57, 74, 93

External gills 9, 10, 19, 20

Facultative air breather 3, 61, 63

Falco peregrinus 102

Fat 1, 131

Faveoli 88, 89, 90-92, 96, 98

Filopodia 83, 117, 128

Fish gills 15, 17, 18, 20, 21, 31, 33, 114, 133

Flying lemur 102

Force 18, 21, 57, 74, 85, 130, 132

Fossorial 70, 73, 117

Fractal 15-17, 133

Freshwater 31, 37, 38, 46, 50, 60, 61, 70, 102

Frog 3, 39, 40, 45, 70, 73, 102

Fuller 5

Gadus morhua 3, 47, 57

Galago senegalensis 103-107, 129

Gallus gallus 3, 101, 109- 112, 114

Gas exchange 3, 7-9, 15, 17, 18, 20, 21, 33, 39, 40, 47, 57, 59, 70-73, 84-86, 92, 93, 102, 113, 114, 116, 130, 132, 133

Gas exchange media 15

Gas exchanger 11

Gas gland 47, 48, 51-56, 58

Gastrointestinal system 9, 65, 69, 130, 132

Geodesic 5

Geometry 14-16, 133

Gill 6, 8-13, 15, 17-21, 30-40, 45, 61, 64, 69, 71, 72, 85, 114, 132-134

Gill arch 22-24, 28, 29

Gill fan 64

Gill filament 17, 21-24, 28, 29, 31, 33, 38, 61

Gill raker 64

Gliding lizard 102, 115, 130

Glucose 2, 57, 131

Glutathione 2

Glycogen 131

Glycolysis 3

Goblet cell 41, 117

Goldfish 3

Golgi bodies 78, 83

Granular cell 84, 95

Guanine crystals 47

Guinea fowl 3, 115

Guinea pig 3

Gymnophiona 20, 70

Gyps rueppellii 102

Habitat 2, 3, 9, 10, 38-40, 46, 61, 70, 73, 85, 101, 116, 133

Harmonic 4, 18, 115

Heart 16, 61, 86, 114, 115, 130

Hematocrit 45, 73, 86, 130, 133

Hemitripterus americanus 3

Hemoglobin 21, 31, 40, 45, 57, 73, 93, 115, 130, 133

Heterocephalus glaber 117

Hirundo fuligula 18, 115

Holostei 60, 61

Homeostasis 2, 134

Homeotherm 1, 9, 86, 101, 116, 132

Homology 9

Honeycomb 5, 117

Hoplostenum thoracatum 57, 69

Hormone 8, 21, 31, 33

Hovering 115, 130

Hummingbird 17, 18, 114, 115

Hydrogen peroxide 2

Hydrogen sulfide 3

Hydrosphere 10

Hydroxyl radical 21

Hyla arborea 73

Hypoxia 15, 45, 46, 61, 114, 130

Icaronycteris index 117

Icefish 21, 40

Ichthyophis paucesulcus 73

Infundibulae 105-107, 112, 114

Insects 102, 115

Inspiration 57, 93

Interalveolar pore 120, 125, 128

Intercostal muscles 85

Interfaveolar septa 93

Internal gills 19, 20, 71

Interparabronchial artery 114

Interparabronchial septa 104

Interstitial cell 21, 75, 98

Interstitial space 8, 74, 117

Intestine 69

Invagination 10-12, 133

Japanese waltzing mouse 117

Keratin 39, 85

Kidney 39, 85

Labyrinthine organ 9, 19, 64-66, 68, 69

Lactic acid 52, 57

Lake Magadi 47, 50

Lake Malawi 69

Lamella 12, 17, 19-21, 26, 31, 33, 38, 40

Land 2, 9, 10, 19, 39, 46

Larus ridibundus 108

Lasionyctris noctivagans 130

Lasiurus borealis 130

Lasiurus cinereus 130

Latimeria chalumnae 130

Lavoisier 1

Lepidosiren paradoxa 10, 20, 61, 63

Lepisosteus osseus 46, 61

Lizard 73, 86, 93, 101, 102

Lobectomy 117

Locomotion 46, 72, 73, 101, 133

Lung-air sac system 16, 17, 93, 101, 102, 130

Lungfish 10, 20, 21, 40, 46, 60, 61, 62, 68, 70, 73, 74, 77, 79, 80, 84, 93, 114, 133

Lungs 1, 7-13, 15-18, 20, 21, 39, 40, 45-47, 60, 61, 63, 64, 70-74, 78, 82, 84-86, 90-94, 96-102, 106, 108, 113, 114-117, 121, 130, 132, 133, 128

Mackerel 21, 33, 38, 47

Macrophage 81-84, 93, 98, 117, 128

Macula adherens 47

Malacochersus tornieri 90, 92, 93

Mammal 1, 3, 8-10, 16-18, 39, 86, 93, 100, 102, 113-117, 130, 133

Mammalian lung 15, 102

Marginal channel 21, 32-34

Megachiroptera 117

Metabolic rate 11, 93, 115, 117

Metabolism 2, 8, 33, 40, 57, 134

Metamorphosis 9, 70, 73

Mezozoic 61, 85

Microchiroptera 117, 130

Microridge 17, 21, 29

Microvilli 53

Miniopterus minor 127, 128

Misgurnus 66

Mitochondria 7, 17, 21, 27, 31, 35-37, 56, 78, 79, 83, 87, 93, 115, 127

Molecule 1, 2, 4

Monitor lizard 86, 91, 98

Mouse 100, 116, 117

Mucus 21, 31, 44, 117

Mudskipper 21

Multicameral 15, 93, 100, 101, 114, 116

Mus musculus 3

Mus wagneri 117

Muscle 8, 10, 17, 20, 33, 47, 57, 59, 73, 85, 86, 93, 97, 98, 100, 105, 115, 130

Mute swan 113

Myoglobin 1

Myotis velifer 117, 130

Naked mole rat 117

Natural selection 4, 5, 9, 131

Nature 3-5, 9, 14-17, 57, 70, 86, 93, 131, 132

Necturus maculatus 20, 47, 72, 73

Neoceratodus 61, 63, 64

Neopulmo 113

Newt 70, 73

Nonpasserine 115

Noradrenaline 31

Numida meleagris 3, 115

Nyctalus noctula 130

Obligate air-breather 10, 61

Onchorhynchus mykiss 3, 33

Oreochromis niloticus 24, 37, 50

Osmiophilic bodies 75, 78, 127

Osmoregulation 8, 31, 33

Osmorespiratory compromise 31

Osteichthyan 19

Osteichthyes 46

Ostrich 18, 101, 113, 115

Oxidative phospholylation 1, 132

Oxidative species 2

Oxygen 1-3, 5-11, 13, 15-20, 31, 33, 39, 40, 45-47, 50, 57, 60, 61, 68, 69, 72, 73, 85, 86, 93, 102, 114-117, 130-133

Oxygen consumption 8, 33, 40, 86, 93, 102, 130

Oxygen extraction 15, 130

Paleopulmo 93, 113, 114

Paleozoic 2, 46, 61, 86

Papio anubis 124-126

Parabronchial lumen 15, 105, 107, 111, 112, 114

Parabronchial lung 101, 114, 130, 132

Parabronchus 104, 107, 114

Parenchyma 10, 93, 100, 113, 114, 117, 119, 130, 132

Partial pressure 2, 6-8, 9, 13, 17, 20, 40, 132

Passerine 113, 115

Paucicameral 86, 92

Pavement cells 21, 27, 85

Paxton 5

Pectoral girdle 85

Pelamis platurus 45

Penguin 113, 115

Perfluocarbon 8

Perfusion 17, 21, 33, 40, 45, 72

Pericyte 78, 99

Perikarya 80

Phanerozoic 2

Photosynthesis 2, 3, 123

Phyllostomus hastatus 18, 102, 123, 130

Physoclistous 47, 57

Physostomous 47, 50, 57, 59

Pig 120

Pigeon 3, 101, 102

Pillar cells 21, 26, 30, 33, 34

Piracucu 60

Plasma 13

Plecostomus 66

Plethodontid 45, 72

Pneumatic duct 57, 59, 113

Pneumocyte 78-80, 84, 93, 95

Polpterus bichir 46, 61, 64

Polypteridae 6, 20, 47

Polytoma 9

Pomatomus saltatrix 21

PreCambrian 2

Pregrine falcon 102

Priestley 1

Primary bronchus 92, 93, 103, 104

Prokaryotic 1

Prostaglandin 8

Protoavis 101

Protopterus aethiopicus 10, 20, 40, 60-64, 76, 77, 79, 80

Protopterus amphibius 64

Protozoa 132

Pseudemys 3, 93

Pteropus gouldii 122

Pterosaurs 102

Pulmonary artery 47, 61

Pulmonary blood flow 16

Pulmonary vasculature 16

Pulmonary vein 61

Rana catesbeiana 40, 73

Rana pipiens 3, 73

Rana temporaria 3, 20

Rattus rattus 3, 20

Reptile 2, 18, 39, 40, 85, 86, 93, 101, 113, 114, 116, 133

Reserve capacity 4

Respiration 1, 5, 8, 10, 33, 39, 40, 45, 47, 60, 69, 70-72, 85

Respiratory design 13

Respiratory fluid 7, 114

Respiratory medium 5, 6, 8-10, 13, 15, 45, 133

Respiratory organs 3, 5, 6, 8-12, 14-21, 46, 50, 60, 61, 64-66, 69-72, 86, 131, 134

Rete mirabile 51, 57

Retina 57

Ruppel's griffon vulture 102

Safety factor 4, 14

Salamanders 20, 40, 45, 47, 70, 72, 73

Salentia 20, 70

Salmo gairdneri 3

Sandelia capensis 69

Sarcopterygian 61

Sarda sarda 21

Scomber scombrus 21

Scyliohinus canicula 3

Sea level 2, 8, 102, 117, 132

Sea raven 3

Secondary bronchi 104

Secondary lamellae 12, 22-26, 28, 30, 32-34

Secretory bodies 51, 54, 55, 58, 79

Secretory cell 44, 65, 66, 68

Secretory pit 53

Septa 71, 72, 76, 77, 81, 88-91, 94, 95, 97, 98, 105, 107, 110, 121, 122, 124

Serotonin 8, 31

Shrew 10, 17, 18, 114, 115, 117

Shunt 20, 21, 61

Siren lacertina 40, 73

Size 5, 14, 16, 18, 20, 21, 40, 46, 86, 93, 101, 102, 113, 115-117, 134

Skeletal muscle 17, 20, 35

Skin 9, 11, 20, 21, 39, 40, 45, 65, 70-73, 85, 132

Smooth muscle 8, 33, 47, 49, 51, 52, 54, 56, 57, 58, 73, 74, 93, 100

Snake 21, 45, 70, 73, 86, 92, 98, 100, 101, 102

Sorex minutus 10

Spheniscus humboldti 115

Sphenodontia 86

Squalus acanthias 3

Squalus suckleyi 3

Squamous cell 17, 21, 47, 84, 117

Starch 131

Sterna paradisea 102

Sternotherus odoratus 40

Sternum 85, 130

Stomach 66, 69

Struthio camelus 18, 101, 113, 115

Sturgeon 17

Sulci 103

Suncus etrucus 18

Superoxide anion 2

Superoxide dismutase 2

Supporting cell 27, 36, 37

Suprabranchial chamber membrane 9, 64-66, 69

Surface 3, 10, 15, 17, 31, 39, 47, 51, 53, 57, 65, 69, 83

Surface area 11, 16, 18, 20, 21, 33, 40, 73, 84, 93, 101, 113-117, 130, 132, 133

Surface tension 18, 31, 113, 117, 130, 132

Surfactant 8, 18, 47, 75, 79, 93, 117, 132

Sus scrofa 120

Swim bladder 9, 46, 47-59, 61, 132

Symmorphosis 17

Syrinx 103

Tadarida brasiliensis 130

Tadarida mops 118

Tadpole 20, 21

Teleost 24, 26-30, 32-37, 43, 44, 48, 49, 51-53

Telmatobius 45, 73

Terrestrial 9-11, 39, 40, 60, 61, 69, 70, 73, 85, 116

Tertiary bronchi 116

Tertiary period 113

Tetrapod 19, 39, 61, 70

Thermodynamics 131

Thoracic cavity 11, 39, 85, 113

Tidal 11, 13, 93, 113, 116, 130, 133

Tissue 1, 2, 4-9, 11, 14, 16, 17, 19, 20, 47, 57, 74, 86, 93, 113-115, 117, 132

Tissue barrier 7, 13

Toad 3, 70, 73

Tortoise 90, 92, 93

Trabeculum 72

Trachea 85, 93, 94, 103, 124

Tracheal system 11

Trachurus mediterraneus 17, 21, 47

Trade-off 18, 33, 39, 133

Transverse capillaries 65-68

Tree frog 41, 42, 45, 70, 71, 73, 75, 78, 81-83

Triassic Period 60

Trionyx mucita 40

Trionyx spiniferous asperus 40

Triturus alpestris 40

Trout 3, 33, 40

Tuna 33, 38

Tupinambis nigropunctus 86, 93

Turtle 3, 40, 85, 86, 93, 101

Type I cell 75, 84, 93, 117, 125, 126

Type II cell 75, 78, 83, 84, 93, 95, 117, 125, 127

Typhlonectes compressicaudata 71, 73

Unicameral 86

Uniform pool 15

Universe 131

Ureotelism 70

Urodela 20, 70, 72

Varanus exanthematicus 91, 93

Vascular channels 23, 25, 26, 30, 34, 35

Vascularization 10, 17, 73

Venous blood 1, 15, 31, 114

Ventilated pool 15

Ventilation 1, 7, 8, 10, 11, 15, 17, 33, 39, 102, 113, 130

Vervet monkey 119, 121, 122

Vesicular bodies 79, 83

Vestigeal 10, 73

Victoria amazonica 5

Viscosity 10, 15, 40, 46, 133

Viviparous 71

Water 2, 5, 6, 8-11, 13, 15, 17-21, 31, 33, 35, 38-40, 45-47, 57, 60, 61, 70-73, 85, 102, 117, 132, 133

Water-blood barrier 7, 20, 21, 69

Water-breather 11

Whale 18

White blood cell 97

Wing beat 115, 130

Xenopus laevis 40, 74

9 781578 082520

T - #0312 - 160425 - C8 - 305/226/9 - PB - 9781578082520 - Gloss Lamination